DISEASES AND THE ORGANISMS THAT CAUSE THEM *(CONTINUED)*

BACTERIAL DISEASES

Disease	Organism	Type*	Page
typhus, scrub (tsutsugamushi disease)	*Rickettsia tsutsugamushi*	CB, NA	684
verruga peruana (bartonellosis)	*Bartonella bacilliformis*	coccoid, −	686
vibriosis	*Vibrio parahaemolyticus*	R, −	643
whooping cough (pertussis)	*Bordetella pertussis*	CB, −	AppB, 606–607
yersiniosis	*Yersinia enterocolitica*	R, −	646–647

*Key to types:

C = coccus
CB = coccobacillus
R = rod
S = spiral
I = irregular
− = Gram-negative
+ = Gram-positive
VAR = Gram-variable
A-F = acid-fast
NA = not applicable

†Species

VIRAL DISEASES

Disease	Virus	Reservoir	Page
aplastic crisis in sickle cell anemia	erythrovirus (B19)	humans	260, 691–692
bronchitis, rhinitis	parainfluenza	humans, some other mammals	600, 601
Burkitt's lymphoma	Epstein-Barr	humans	689
chickenpox	varicella-zoster	humans	261, 265, 544–546
coryza (common cold)	rhinovirus	humans	260, 262,
	coronavirus	humans	603–604
cytomegalic inclusion disease	cytomegalovirus	humans	590–591
Dengue fever	Dengue	humans	314, 667–668
encephalitis	Colorado tick fever	mammals	314, 691
	Eastern equine encephalitis	birds	262–263, 405, 706–707
	St. Louis encephalitis	birds	706–707
	Venezuelan equine encephalitis	rodents	262–263, 706–707
	Western equine encephalitis	birds	262–263, 314, 405, 706–707
epidemic keratoconjunctivitis	adenovirus	humans	555
fifth disease (erythema infectiosum)	erythrovirus (B19)	humans	265–267, 692
hantavirus pulmonary syndrome	bunyavirus	rodents	263, 622
hemorrhagic fever, Bolivian	arenavirus	rodents and humans	691
hemorrhagic fever, Korean	bunyavirus (Hantaan)	rodents	263, 691
hemorrhagic fever	Ebola virus (filovirus)	humans (?)	260, 263, 690
	Marburg virus (filovirus)	humans (?)	260, 263, 690
hepatitis A (infectious hepatitis)	hepatitis A	humans	260, 651–652
hepatitis B (serum hepatitis)	hepatitis B	humans	261, 651–653
hepatitis C (non-A, non-B)	hepatitis C	humans	651–653
hepatitis D (delta hepatitis)	hepatitis D	humans	651–653
hepatitis E (enterically transmitted non-A, non-B, non-C)	hepatitis E	humans	651–653
herpes, oral	usually herpes simplex type 1, sometimes type 2	humans	261, 264, 589
herpes, genital	usually herpes simplex type 2, sometimes type 1	humans	261, 264, 588
HIV disease, AIDS	human immunodeficiency virus (HIV)	humans	260, 518–523
infectious mononucleosis	Epstein-Barr	humans	688–689
influenza	influenza	swine, humans (type A)	260, 263 616–621
		humans (typeB)	260, 263 616–621
		humans (type (type C)	616–621
Lassa fever	arenavirus	rodents	691
measles (rubeola)	measles	humans	260, 543–544
meningoencephalitis	herpes	humans	589, 707
molluscum contagiosum	poxvirus group	humans	547
mumps	paramyxovirus	humans	637
pneumonia	adenoviruses, respiratory syncytial virus	humans	AppB, 608–609
poliomyelitis	poliovirus	humans	260, 713–715
rabies	rabies	all warm-blooded animals	260, 263, 416 704–706
respiratory infections	adenovirus	humans	622
	polyomavirus	none	707–708
Rift Valley fever	bunyavirus (phlebovirus)	humans, sheep, cattle	691
rubella (German measles)	rubella	humans	260, 541–543
shingles	varicella-zoster	humans	261, 264, 544
smallpox	variola (major and minor)	humans	261, 265, 546–547
viral enteritis	rotavirus	humans	649–651
warts, common (papillomas)	human papillomavirus	humans	261, 547–549
warts, genital (condylomas)	human papillomavirus	humans	261, 547–548
yellow fever	yellow fever	monkeys, humans, mosquitoes	260, 263, 266 314, 688

The tables of fungal and parasitic diseases appear on the following page.

DISEASES AND THE ORGANISMS THAT CAUSE THEM *(CONCLUDED)*

FUNGAL DISEASES

Disease	Organism	Page
aspergillosis	*Aspergillus* sp.	551–2, 625
blastomycosis	*Blastomyces dermatitidis*	550–1
candidiasis	*Candida albicans*	551
coccidioidomycosis (San Joaquin valley fever)	*Coccidioides immitis*	623–4
cryptococcosis	*Filobasidiella neoformans*	624
ergot poisoning	*Claviceps purpurea*	757
histoplasmosis	*Histoplasma capsulatum*	624
Pneumocystis pneumonia	*Pneumocystis carinii*	625
ringworm (tinea)	various species of *Epidermophyton, Trichophyton, Microsporum*	549–50
sporotrichosis	*Sporothrix schenckii*	550
zygomycosis	*Rhizopus* sp., *Mucor* sp.	552

PARASITIC DISEASES

Disease	Organism	Type	Page
Acanthamoeba keratitis	*Acanthamoeba culbertsoni*	protozoan	413
African sleeping sickness (trypanosomiasis)	*Trypanosoma brucei gambiense* and *T. brucei rhodesiense*	protozoan	294, 313, 314 369, 718–9
amoebic dysentery	*Entamoeba histolytica*	protozoan	291,294, 655
ascariasis	*Ascaris lumbricoides*	roundworm	660–1
babesiosis	*Babesia microti*	protozoan	696–8
balantidiasis	*Balantidium coli*	protozoan	655–6
Chagas' disease	*Trypanosoma cruzi*	protozoan	314, 383 369, 720–1
chigger dermatitis	*Trombicula* sp.	mite	560
chigger infestation	*Tunga penetrans*	sandflea	560
Chinese liver fluke	*Clonorchis sinensis*	flatworm	658
crab louse	*Phthirus pubis*	louse	561
cryptosporidiosis	*Cryptosporidium* sp.	protozoan	656
dracunculiasis (Guinea worm)	*Dracunculus medinensis*	roundworm	552
elephantiasis (filariasis)	*Wuchereria bancrofti*	roundworm	308, 309 672, 673
fasciolopsiasis	*Fasciolopsis buski*	flatworm	658
giardiasis	*Giardia intestinalis*	protozoan	654–5
heartworm disease	*Dirofilaria immitis*	roundworm	290, 305, 369
hookworm	*Ancylostoma duodenale* (Old World hook-worm)	roundworm	305, 308, 660
	Necator americanus (New World hook-worm)	roundworm	308, 660
leishmaniasis	*Leishmania braziliensis*	protozoan	291,313, 314, 692–3
kala azar	*L. donovani*		
oriental sore	*L. tropica*		
liver/lung fluke (paragonimiasis)	*Paragonimus westermani*	flatworm	306, 657, 658
loaiasis	*Loa loa*	roundworm	313, 556–7
malaria	*Plasmodium* sp.	protozoan	71, 291, 294–5, 418–20, 693–96,
pediculosis (lice infestation)	*Pediculus humanus*	louse	560–1
pinworm	*Enterobius vermicularis*	roundworm	369, 376, 662
river blindness (onchocerciasis)	*Onchocerca volvulus*	roundworm	369, 556
scabies (sarcoptic mange)	*Sarcoptes scabiei*	mite	560
schistosomiasis	*Schistosoma* sp.	flatworm	306, 552, 670–2
sheep liver fluke (fascioliasis)	*Fasciola hepatica*	flatworm	657
strongyloidiasis	*Strongyloides stercoralis*	roundworm	662
swimmer's itch	*Schistosoma* sp.	flatworm	552
tapeworm infestation (taeniasis)	*Hymenolepis nana* (dwarf tapeworm)	flatworm	658–9
	Taenia saginata (beef tapeworm)	flatworm	658–9
	Taenia solium (pork tapeworm)	flatworm	658–9
	Diphyllobothrium latum (fish tapeworm)	flatworm	658–9
	Echinococcus granulosus (dog tapeworm)	flatworm	658–9
toxoplasmosis	*Toxoplasma gondii*	protozoan	291, 295, 696
trichinosis	*Trichinella spiralis*	roundworm	308, 334, 659–60
trichomoniasis	*Trichomonas vaginalis*	protozoan	573–4
trichuriasis (whipworm)	*Trichuris trichiura*	roundworm	661–2
visceral larva migrans	*Toxocara* sp.	roundworm	661

Microbiology

Jacquelyn G. Black

MARYMOUNT UNIVERSITY, ARLINGTON, VIRGINIA

Jacquelyn Black received her B.A., B.S., and M.S. from the University of Chicago and her Ph.D. from Catholic University of America. She has been teaching microbiology to undergraduates since 1970. She is a member of the American Society for Microbiology, and she has received grants for conducting teacher-training programs.

In addition to her extensive teaching experience, Dr. Black has engaged in fieldwork and studies throughout the globe. Her travels have taken her from the interior of Iceland to Belgium and Portugal to the barrier reef of Belize.

Dr. Black describes herself as an "incorrigible snoop" who is interested in all the various aspects and applications of microbiology. This natural curiosity, coupled with her classroom and laboratory experience, make her uniquely qualified to author an introductory microbiology textbook. This book conveys her sense of excitement for microbiology and offers the most current information on developments and applications within this field.

Dedication
To Laura—for sharing her mother and much of her childhood with "the book."

Microbiology

Principles and Explorations

Jacquelyn G. Black

Marymount University

JOHN WILEY & SONS, INC.

New York • Chichester • Weinheim • Brisbane • Singapore • Toronto

Scope and History of Microbiology

When you think about microbiology, you might think about nasty little organisms that are so small they are invisible to the naked eye. You might think about petri dishes and microscopes and people in white lab coats. Perhaps you think about germs and you hear your mother's voice calling from your childhood, "Don't put that in your mouth. You don't know where it's been, and it has germs all over it!" When you think about microbiology you probably DON'T think about our planet Earth revolving in space. But you should.

Without microbes Earth as we know it would not exist, and you would not exist. Microbes are relevant to virtually all aspects of human life. You cannot eat a hamburger, sit in a hot tub, or even brush your teeth without encountering microbes. They are everywhere, on you and around you. They're even on this page. Furthermore, microbes are not all bad; many of them actually improve the quality of life on Earth.

So next time you look at a photo of our blue planet, celebrate the fact that it has "germs" (and other microbes) all over it. And by the way, you might want to mention this to your mother.

Why Study Microbiology?

Microbes (bacteria, viruses, fungi, protozoa, and some algae) live in us, on us, and nearly everywhere around us (➢ Figure 1.1). They have a major impact on our health and environment. They play an important role in many of the foods we eat and the medicines we take. They also provide useful models for many of the life processes that all organisms experience.

Microbes Have a Major Impact on Human Health

Though fewer than 1 percent of all microbes cause disease, identifying how those disease-causing microbes (called pathogens) cause and transmit disease, as well as discovering how those diseases can be treated, is of critical importance to our well-being as a species. In 1962, the U.S. Surgeon General, William H. Stewart, emphatically stated, "The war against infectious diseases has been won." In the subsequent decades, emerging diseases and increasing antibiotic resistance have made it clear that in reality not only is the war far from won, its outcome is no longer certain. It will be our health science professionals who will be on the frontlines of that war into the foreseeable future.

Microbes Help Maintain the Balance of Nature

Many aquatic microorganisms capture energy from sunlight and store it in molecules that other oganisms use as food. Microorganisms decompose dead organisms and waste material from living organisms, and they can decompose some kinds of industrial wastes. Through this decomposition, they make nitrogen available to plants. Certain microorganisms that reside in the digestive tracts of grazing animals play an important role in those animals' ability to digest grass.

Questions We'll Explore

A Why is the study of microbiology important?

B What is the scope of microbiology?

C What are some major events in the early history of microbiology?

D What is the germ theory of disease, and what historical developments led to its formulation?

E What events mark the emergence of immunology, virology, chemotherapy, genetics, and molecular biology as branches of microbiology?

Almost one-half of children under the age of 10 died of infectious disease prior to this century.

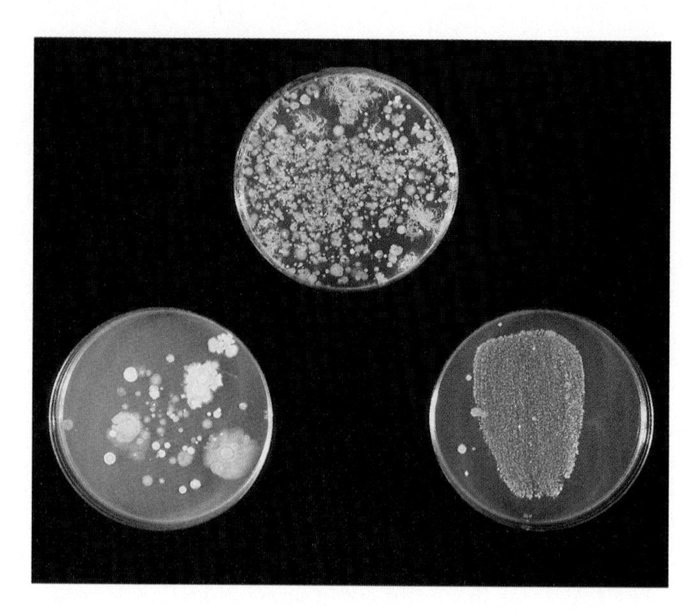

➢ Figure 1.1 **A simple experiment shows that microorganisms are almost everywhere in our environment.** Soil was added to nutrient agar, a culture medium (dish on top); another dish with agar was exposed to air (bottom left); and a tongue print was made on an agar surface (bottom right). After 3 days of incubation under favorable conditions, abundant microbial growth is easily visible in all three dishes.

These are just a few examples of the integral nature of microorganisms in our environment. The vast majority of microbes are directly or indirectly beneficial to other organisms, humans included. They form essential links in many food chains that produce the plans and animals that people eat.

Microorganisms Are Essential to Many Human Endeavors

The biochemical reactions carried out by microorganisms have been harnessed by humans for our own ends. The food industry uses these reactions in the preparations of pickles, sauerkraut, yogurt, fructose, and aspartame, the artificial sweetener. Fermentation reactions are exploited in the creation of beer and wine, as well as in the baking of bread. One of the most significant benefits that microorganisms provide is their ability to synthesize **antibiotics,** substances derived from one organism that kill or inhibit the growth of other microorganisms.

These are all contributions that microorganisms make in their natural state. The advent of genetic engineering has created a whole new suite of benefits. Several substances important to humans, such as interferon and growth hormones, can be economically produced by genetically altered microbes. New organisms have been engineered to degrade oil spills, to remove toxic materials from the soil, and to digest explosives. As we move into the future, it is safe to say that genetically engineered microorganisms will play an increasing role in many aspects of human affairs whether they be medical, environmental, or economic.

Microbiology Provides Insight Into Life Processes in All Life Forms

Biologists in many different disciplines use both the ideas from microbiology and the organisms themselves. Ecologists draw on principles of microbiology to understand how matter is decomposed and made available for continuous recycling. Biochemists use microbes to study metabolic pathways—sequences of chemical reactions in living organisms. Geneticists study microorganisms to investigate the processes of gene transmission.

Microorganisms are particularly attractive to many researchers because:

1. Compared to other organisms, microbes have relatively simple structures. It is easier to study most life processes in simple unicellular organisms than in complex multicellular ones.
2. Large numbers of microorganisms can be used in an experiment to obtain statistically reliable results at a reasonable cost. Growing a billion bacteria costs less than maintaining 10 rats. Experiments with large numbers of microorganisms give more reliable results than do those with small numbers of organisms with individual variations.

A bacterium may weigh approximately 0.00000000001 gram.

3. Because microorganisms reproduce very quickly, they are epecially useful for studies involving the transfer of genetic information. Some bacteria can undergo three cell divisions in an hour, so the effects of gene transfer can quickly be followed through many generations.

By studying microbes, scientists have achieved remarkable success in understanding life processes and disease control. For example, within the last few decades, vaccines have nearly eradicated several dreaded childhood diseases—including measles, polio, German measles, and mumps. Recently a vaccine was developed for chickenpox. Smallpox, which once accounted for 1 out of every 10 deaths in Europe, has not been reported anywhere on the planet since 1978. Much has also been learned about genetic changes that lead to antibiotic resistance and about how to manipulate genetic information in bacteria. Much more remains to be learned. For examples, how can vaccines be made available on a worldwide basis? How can the development of new antibiotics keep pace with genetic changes in microorganisms? How will increased jet-age world travel affect the spread of infections? Will the continued encroachment of humans into virgin forests result in new emerging diseases? Can a vaccine or effective treatment for AIDS be developed? Therein lie some of the challenges for the next generation of biologists and health scientists.

Scope of Microbiology

Microbiology is the study of **microbes,** organisms so small that a microscope is needed to study them. We consider two dimensions of the scope of microbiology: (1) the variety of kinds of microbes and (2) the kinds of work microbiologists do.

The Microbes

The major groups of organisms studied in microbiology are bacteria, algae, fungi, viruses, and protozoa (➢ Figure 1.2a-e). All are widely distributed in nature. For example, a recent study of bee bread (a pollen-derived nutrient eaten by worker bees) showed it to contain 188 kinds of fungi and 29 kinds of bacteria. Most microbes consist of a single cell. (Cells are the basic units of structure and function in living things; they are discussed in Chapter 4.) Viruses, tiny acellular entities on the borderline between the living and the nonliving, behave like living organisms when they gain entry to cells. They, too, are studied in microbiology. Microbes range in size from small viruses 20 nm in diameter to large protozoans 5 mm or more in diameter. In other words, the largest microbes are as much as 250,000 times the size of the smallest ones! (Refer to Appendix A for a review of metric units.)

Among the great variety of microorganisms that have been identified, bacteria probably have been the most thoroughly studied. The majority of **bacteria** (singular: *bacterium*) are single-celled organisms with spherical, rod, or spiral shapes, but a few types form filaments. Most are so small they can be seen with a light microscope only under very high magnification. Although bacteria are cellular, they do not have a cell nucleus, and they lack the membrane-enclosed intracellular structures found in most other cells. Many bacteria absorb nutrients from their environment, but some make their own nutrients by photosynthesis or other synthetic processes. Some are stationary, and others move about. Bacteria are widely distributed in nature, for example, in aquatic environments and in decaying matter. And some occasionally cause diseases.

In contrast to bacteria, several groups of microorganisms consist of larger, more complex cells that have a cell nucleus. They include algae, fungi, and protozoa, all of which can easily be seen with a light microscope.

Many **algae** (al'je; singular: *alga*) are single-celled microscopic organisms, but some marine algae are large, relatively complex, multicellular organisms. Unlike bacteria, algae have a clearly defined cell nucleus and numerous membrane-enclosed intracellular structures. All algae photosynthesize their own food as plants do, and many can move about. Algae are widely distributed in both fresh water and oceans. Because they are so numerous and because they capture energy from sunlight in the food they make, algae are an important source of food for other organisms. Algae are of little medical importance; only one species has been found to cause disease in humans.

We Are Not Alone

"We are outnumbered. The average human contains about 10 trillion cells. On that average human are about 10 times as many microorganisms, or 100 trillion microscopic beings. . . . As long as they stay in balance and where they belong, [they] do us no harm. . . . In fact, many of them provide some important services to us. [But] most are opportunists, who if given the opportunity of increasing growth or invading new territory, will cause infection."

—Robert J. Sullivan, 1989

500 bacteria, each 1 μm (1/1000 of a millimeter) long, would fit end-to-end across the dot above the letter "i."

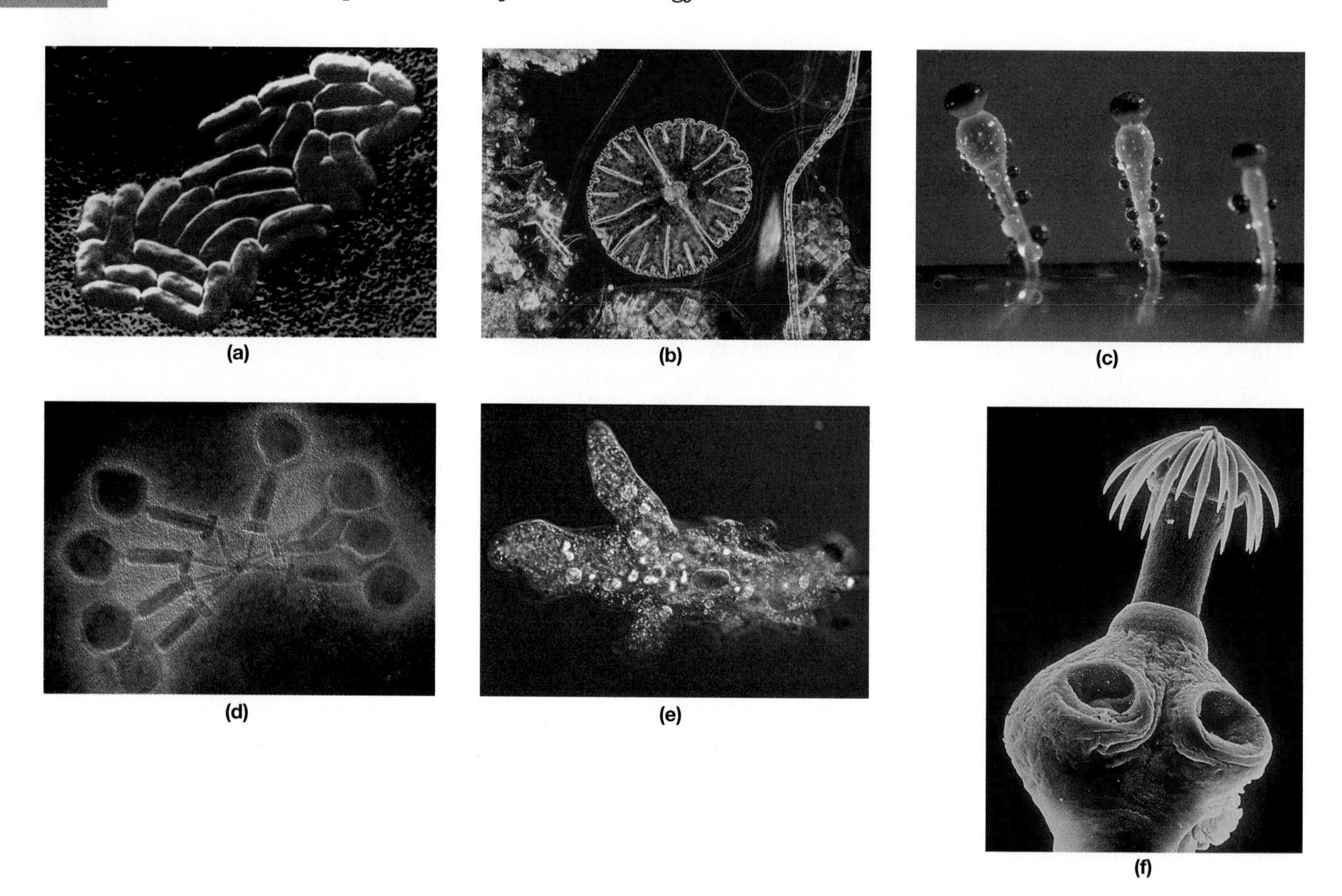

➢ Figure 1.2 **Typical microorganisms.** (a) Several *Klebsiella pneumoniae* cells (magnified 1120X), a bacterium that can cause pneumonia in humans. (b) *Micrasterias,* a type of green algae that lives in fresh water. (c) Fruiting bodies of the fungus *Philobolus crystallinus* with black spore packets on top that will be shot into the air to colonize new areas. (d) Bacteriophages (viruses that infect bacteria; 35,500X). (e) *Amoeba,* a protozoan (175X). (f) Head of the tapeworm *Acanthrocirrus retrirostris* (170X). At the top of the head are hooks and suckers that the worm uses to attach to a host's intestinal tissues.

Like algae, many **fungi** (fun′ji; singular: *fungus*), such as yeasts and some molds, are single-celled microscopic organisms. Some, such as mushrooms, are multicellular, macroscopic organisms. Fungi also have a cell nucleus and intracellular structures. All fungi absorb ready-made nutrients from their environment. Some fungi form extensive networks of branching filaments, but the organisms themselves generally do not move. Fungi are widely distributed in water and soil as decomposers of dead organisms. Some are important in medicine either as agents of disease or as sources of antibiotics.

Viruses are acellular entities too small to be seen with a light microscope. They are composed of specific chemical substances—a nucleic acid and a few proteins (Chapter 2). Indeed, some viruses can be crystallized and stored, but they retain the ability to invade cells. Viruses replicate themselves and display other properties of living organisms only when they have invaded cells. Many viruses can invade human cells and cause disease.

Protozoa (pro-to-zo′ah; singular: *protozoan*) also are single-celled, microscopic organisms with at least one nucleus and numerous intracellular structures. A few species of amoebae are large enough to be seen with the naked eye, but we can study their structure only with a microscope. Many protozoa obtain food by engulfing or ingesting smaller microorganisms. Most protozoa can move, but a few, especially those that cause human disease, cannot. Protozoa are found in a variety of water and soil environments.

In addition to organisms properly in the domain of microbiology, in this text we consider some macroscopic *helminths* (worms) (➢ Figure 1.2f) and *arthropods* (insects and similar organisms). The helminths have microscopic stages in their life cycles that can cause disease, and the arthropods transmit these stages.

We will learn more about the classification of microorganisms in Chapter 9. For now it is important to know only that cellular organisms are referred to by two names: their *genus* and *species* names. For example, a bacterial species commonly found in the human gut is called *Escherichia coli,* and a protozoan species that can cause severe diarrhea is called *Giardia intestinalis*. The naming of viruses is less precise. Some viruses, such as herpesviruses, are named for the group to which they belong. Others, such as polioviruses, are named for the disease they cause.

Disease-causing organisms and the diseases they cause in humans are discussed in detail in Chapters 19–24. Hundreds of infectious diseases are known to medical science. Some of the most important—those diseases that physicians should report to the U.S. Centers for Disease Control and Prevention (CDC)—are listed in Table 1.1 according to the kind of causative organism. The CDC is a federal agency that collects data about diseases and about developing ways to control them.

The Microbiologists

Microbiologists study many kinds of problems that involve microbes. Some study microbes mainly to find out more about a particular type of organism—the life stages of a particular fungus, for example. Other microbiologists are interested in a particular kind of function, such as the metabolism of a certain sugar or the action of a specific gene. Still others focus directly on practical problems, such as how to purify or synthesize a new antibiotic or how to make a vaccine against a particular disease. Quite often the findings from one project are useful in another, as when agricultural scientists use information from microbiologists to control pests and improve crop yields, or when environmentalists attempt to maintain natural food chains and prevent damage to the environment. Some fields of microbiology are described in Table 1.2.

➢ Table 1.1 **Reportable diseases caused by microorganisms and parasites**[a]

Bacterial diseases	Bacterial diseases	Bacterial diseases	Viral diseases	Algal diseases
Anthrax	Hemolytic uremic syndrome	Rocky Mountain spotted fever	AIDS (symptomatic cases)	None
Bacterial meningitis	Legionnaires' disease	Salmonellosis	Arbovirus infection	**Fungal Diseases**
Botulism	Leprosy (Hansen's disease)	Shigellosis	Encephalitis	None
Brucellosis	Listeriosis	Streptococcal disease, invasive, Group A	Hantavirus pulmonary syndrome	**Protozoan Diseases**
Campylobacteriosis	Lyme disease	Streptococcal pneumonia, drug-resistant invasive disease	Hepatitis A, B, and C	Giardiasis
Chancroid	Meningitis	Syphilis	Hepatitis (unspecified)	Malaria
Chlamydial infections	Paratyphoid fever A, B, and C	Tetanus	HIV infection, adult	**Helminth Diseases**
Cholera	Pertussis (whooping cough)	Toxic shock syndrome	HIV infection, pediatric	Trichinosis
Coccidiomycosis	Plague	Trachoma	Influenza	
Cryptosporidiosis	Psittacosis	Tuberculosis	Measles (rubeola)	
Diphtheria	Q fever	Typhoid fever	Mumps	
Food poisoning	Relapsing fever	Typhus	Poliomyelitis	
Gonorrhea			Rabies (animal and human)	
Haemophilus influenzae infections (invasive)			Rubella (German measles)	
			Yellow fever	

[a]Infectious-disease reporting varies by state. This table lists most of the diseases commonly reported to the U.S. Centers for Disease Control and Prevention.

Table 1.2 Fields of microbiology

Field (pronunciation)	Examples of What Is Studied
Microbial taxonomy	Classification of microorganisms
Fields according to organisms studied	
Bacteriology (bak″ter-e-ol′o-je)	Bacteria
Phycology (fi-kol′o-je)	Algae (*phyco,* seaweed)
Mycology (mi-kol′o-je)	Fungi (*myco,* a fungus)
Protozoology (pro″to-zo-ol′o-je)	Protozoa (*proto,* first; *zoo,* animal)
Parasitology (par″a-si-tol′o-je)	Parasites
Virology (vi-rol′o-je)	Viruses
Fields according to processes or functions studied	
Microbial metabolism	Chemical reactions that occur in microbes
Microbial genetics	Transmission and action of genetic information in microorganisms
Microbial ecology	Relationships of microbes with each other and with the environment
Health-related fields	
Immunology (im"u-nol'o-je)	How host organisms defend themselves against microbial infection
Epidemiology (ep-i-de-me-ol'o-je)	Frequency and distribution of diseases
Etiology (e-te-ol'-o-je)	Causes of disease
Infection control	How to control the spread of nosocomial (nos-o-ko'me-al), or hospital-acquired, infections
Chemotherapy	The development and use of chemical substances to treat diseases
Fields according to applications of knowledge	
Food and beverage technology	How to protect humans from disease organisms in fresh and preserved foods
Environmental microbiology	How to maintain safe drinking water, dispose of wastes, and control environmental pollution
Industrial microbiology	How to apply knowledge of microorganisms to the manufacture of fermented foods and other products of microorganisms
Pharmaceutical microbiology	How to manufacture antibiotics, vaccines, and other health products
Genetic engineering	How to use microorganisms to synthesize products useful to humans

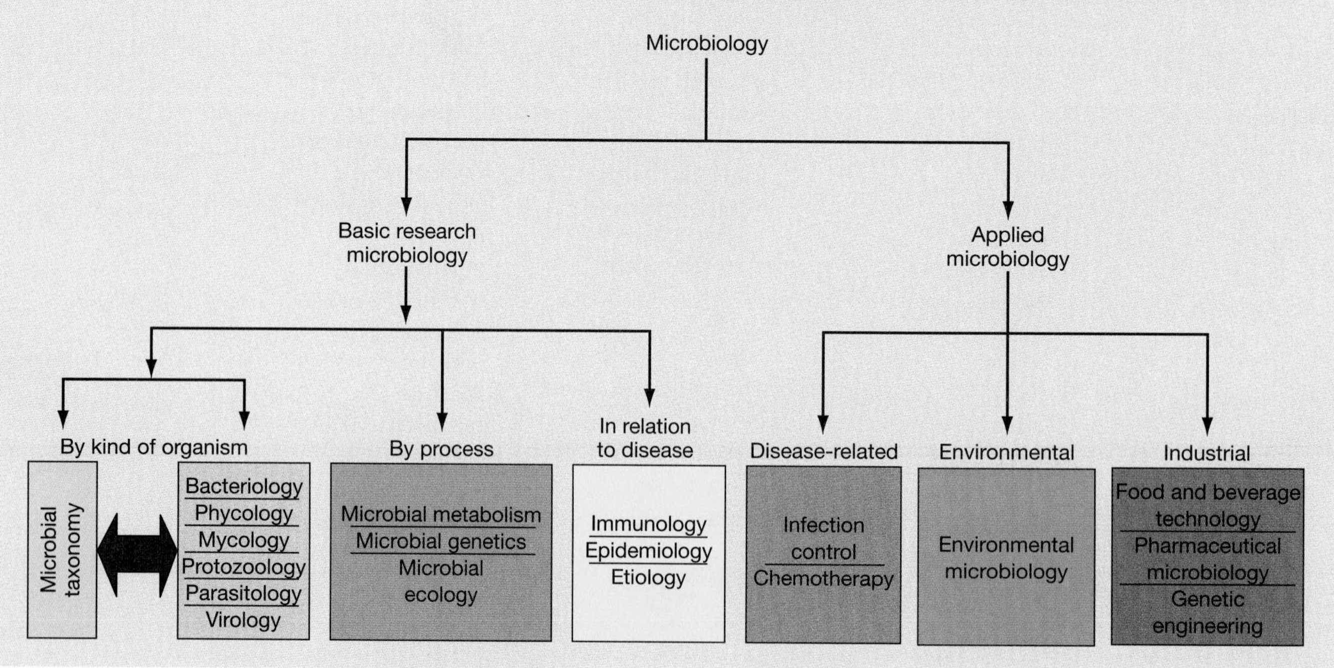

Microbiologists work in a variety of settings (➢ Figure 1.3). Some work in universities, where they are likely to teach, do research, and train students to do research. Microbiologists in both university and commercial laboratories are helping to develop the microorganisms used in genetic engineering. Some law firms are hiring microbiologists to help with the complexities of patenting new genetically engineered organisms. These organisms can be used in such important ways as cleaning up the environment (*bioremediation*), controlling insect pests, improving foods, and fighting disease. Many microbiologists work in health-related positions. Some work in clinical laboratories, performing tests to diagnose diseases or determining which antibiotics will cure a particular disease. A few microbiologists develop new clinical tests. Others work in industrial laboratories to develop or manufacture antibiotics, vaccines, or similar biological products. Still others, concerned with controlling the spread of infections and related public health matters, work in hospitals or government labs.

From the point of view of health scientists, today's research is the source of tomorrow's new technologies. Research in *immunology* is greatly increasing our knowledge of how microbes trigger host responses and how the microbes escape these responses. It also is contributing to the development of new vaccines and to the treatment of immunologic disorders. Research in *virology* is improving our understanding of how viruses cause infections and how they are involved in cancer. Research in *chemotherapy* is increasing the number of drugs available to treat infections and is also improving our knowledge of how these drugs work. Finally, research in

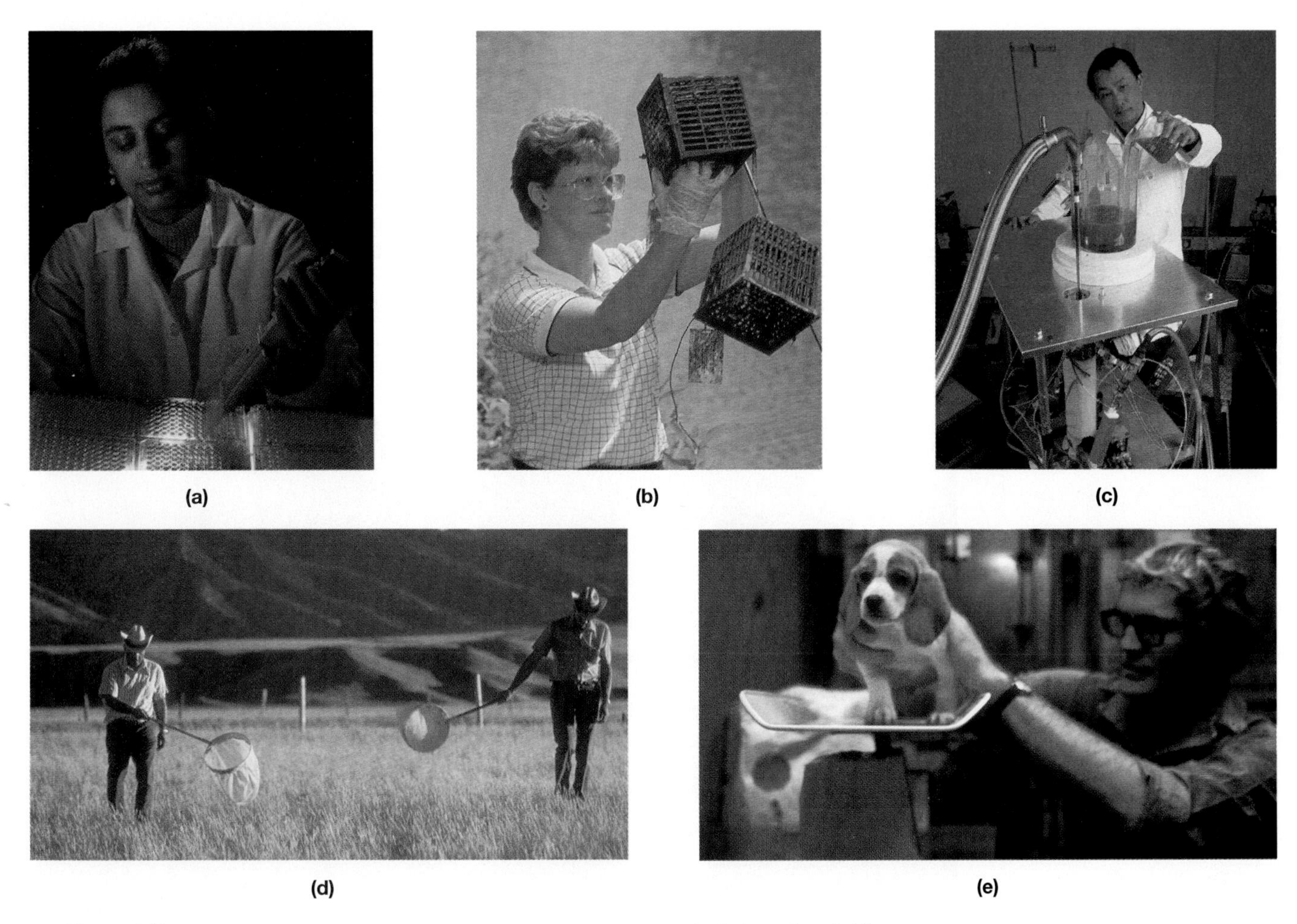

➢ Figure 1.3 **Microbiology is used in diverse careers.** These careers include such activities as (a) using genetically engineered bacteria to investigate how diet influences the risk of developing cancer; (b) inspecting plastics made with as much as 40 percent starch (pieces inside baskets) for signs that aquatic microbes are degrading them; (c) using bacteria to decontaminate toxic wastes; (d) using beating nets to survey for ticks that can spread disease to livestock and humans; (e) keeping pets and domestic animals healthy, as well as improving their productivity, by means of advances in veterinary science.

✓ List three reasons to study microbiology.

✓ What is the difference between microbiology and bacteriology?

✓ What is the difference between etiology and epidemiology?

✓ List 5 bacterial diseases; 5 viral diseases.

genetics is providing new information about the transfer of genetic information and, especially, about how genetic information acts at the molecular level.

Historical Roots

Many of the ancient Mosaic laws found in the Bible about basic sanitation have been used through the centuries and still contribute to our practices of preventive medicine. In Deuteronomy, Chapter 13, Moses instructed the soldiers to carry spades and bury solid waste matter. The Bible also refers to leprosy and to the isolation of lepers. Although in those days the term *leprosy* probably included other infectious and noninfectious diseases, isolation did limit the spread of the infectious diseases.

The Greeks anticipated microbiology, as they did so many things. The Greek physician Hippocrates, who lived around 400 B.C., set forth ethical standards for the practice of medicine that are still in use today. Hippocrates was wise in human relations and also a shrewd observer. He associated particular signs and symptoms with certain illnesses and realized that diseases could be transmitted from one person to another by clothing or other objects. At about the same time, the Greek historian Thucydides observed that people who had recovered from the plague could take care of plague victims without danger of getting the disease again.

The Romans also contributed to microbiology, as early as the first century B.C. The scholar and writer Varro proposed that tiny invisible animals entered the body through the mouth and nose to cause disease. Lucretius, a philosophical poet, cited "seeds" of disease in his *De Rerum Natura (On the Nature of Things).*

Bubonic plague, also called the Black Death, appeared in the Mediterranean region around A.D. 542, where it reached epidemic proportions and killed millions. In 1347 the plague invaded Europe along the caravan routes and sea lanes from central Asia, affecting Italy first, then France, England, and finally northern Europe. Although no accurate records were kept at that time, it is estimated that tens of millions of people in Europe died during this and successive waves of plague over the next 300 years. The Black Death was a great leveler—it killed rich and poor alike (➢ Figure 1.4). The wealthy fled to isolated summer homes but carried plague-infected fleas with them in unwashed hair and clothing.

One group that escaped the plague's devastation was the Jewish population. Ancient Hebrew laws regarding sanitation offered some protection to those who practiced them. The relatively clean Jewish ghettos harbored fewer rats to spread the disease. When Jews did fall ill,

➢ Figure 1.4 **A portion of "The Triumph of Death" by Pieter Brueghel the Elder.** The picture, painted in the mid-sixteenth century, a time when outbreaks of plague were still common in many parts of Europe, dramatizes the swiftness and inescapability of death for people of all social and economic classes.

they were carefully nursed and treated with herbal remedies rather than by strenuous purging or excessive bleedings with dirty instruments. As a result, a smaller proportion of Jews than gentiles died of the disease. Ironically, some gentiles regarded the Jews' higher survival rates as proof that Jews were the source of the epidemic.

In his *Diary,* the English writer Samuel Pepys gave a vivid, firsthand account of the plague in London in the 1660s.

> The streets mighty empty all the way now even in London, which is a sad sight. . . . Poor Will, that used to sell us ale, . . . his wife and three children died, all I think in a day. . . . [H]ome to draw over anew my will, which I had bound myself by oath to dispatch by tomorrow night, the town growing so unhealthy that a man cannot depend upon living two days to an end. In the City died this week 7,496, and of them 6,102 of the plague. But it is feared that the true number of the dead is near 10,000; partly from the poor that cannot be taken notice of through the greatness of the number. . . . I saw a dead corps in a coffin lie in the Close unburied; and a watch is constantly kept there night and day to keep the people in, the plague making us cruel as doggs one to another.

Until the seventeenth century, the advance of microbiology was hampered by the lack of appropriate tools to observe microbes. Around 1665, the English scientist Robert Hooke built a compound microscope (one in which light passes through two lenses) and used it to observe thin slices of cork. He coined the term *cell* to describe the orderly arrangement of small boxes that he saw because they reminded him of the cells (small, bare rooms) of monks. However, it was Anton van Leeuwenhoek (➢ Figure 1.5), a Dutch cloth merchant and amateur lens grinder, who first made and used lenses to observe living microorganisms. The lenses Leeuwenhoek made were of excellent quality; some gave magnifications up to 300× and were remarkably free of distortion. Making these lenses and looking through them were the passions of his life. Everywhere he looked he found what he called "animalcules." He found them in stagnant water, in sick people, and even in his own mouth.

Over the years Leeuwenhoek observed all the major kinds of microorganisms—protozoa, algae, yeast, fungi, and bacteria in spherical, rod, and spiral forms. He once wrote, "For my part I judge, from myself (howbeit I clean my mouth like I've already said), that all the people living in our United Netherlands are not as many as the living animals that I carry in my own mouth this very day." Starting in the 1670s he wrote numerous letters to the Royal Society in

➢ Figure 1.5 **Anton van Leeuwenhoek (1632–1723).** He is shown holding one of his simple microscopes.

AIDS

Suppose that several close relatives, two of your best friends, and many of your neighbors are suffering from painful illnesses and will soon die. There are no available beds at the hospital, and most of the doctors and nurses have quit work because they fear becoming infected. The local television station has started broadcasting daily the latest figures on deaths and new outbreaks, like the stock market prices or the weather report. Nearly every time you go shopping, you meet a funeral procession.

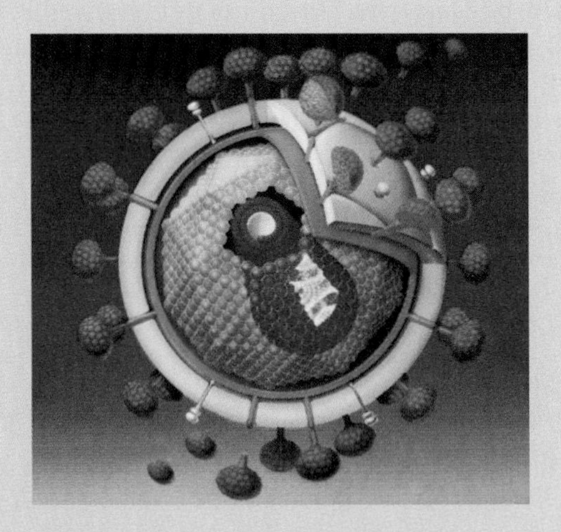

A model of the virus that causes AIDS.

Over the centuries, many people have found themselves in situations that are similar to this fictional modern epidemic. Again and again, large proportions of the human population have been devastated by infectious diseases such as typhus, smallpox, and bubonic plague. In the mid-fourteenth century (1347–1351), plague alone wiped out 25 million people—one-fourth the population of Europe and neighboring regions—in just 5 years.

We who live in technologically advanced countries tend to think that outbreaks of this sort are a thing of the past. Yet following World War I, within the memory of many people now living, worldwide outbreaks of influenza claimed 20 million lives. Today, acquired immune deficiency syndrome (AIDS) threatens to kill great numbers of people after they have suffered a long and painful illness. Is AIDS in any way comparable to the great killer diseases of the past? More alarmingly, does its appearance mean that all our "triumphs" over infectious disease were an illusion? Will the past few decades, during which vaccines and antibiotics have largely kept contagious disease in check and even have wiped out certain ancient curses such as smallpox, prove to have been just a brief and atypical episode in an endless war that can't be won?

The AIDS Name Project, a memorial to victims of this disease, showing individual quilts on display in Washington, D.C.

Despite the fears AIDS has aroused, there are many differences between AIDS and the epidemic diseases mentioned above. For one thing, AIDS is not nearly so easily communicable as the epidemic killers of the past. It is not spread by casual contact but largely by certain behaviors, most of which people can learn to avoid (as we will see in Chapter 19).

More important, however, is the state of our knowledge. During past epidemics, people were terrified and demoralized because they had no idea how the disease was caused and spread or how to fight it. Today, we understand far more about the nature of the enemy. The study of AIDS has occupied virologists, chemotherapists, and immunologists from among the world's most talented microbiologists. They have shown that AIDS is caused by a virus that infects cells of the body's immune system. The virus alters genetic information so that instead of fighting the infection, infected cells make viruses and then die. Scientists have determined the precise structure of many of the components of the virus, and they have learned how it attaches itself to its target cells. They have developed some drugs that are being tested to treat AIDS, and they are working on a vaccine. Despite those efforts, however, AIDS remains one of the most threatening infectious diseases and one of the greatest challenges microbiologists have ever faced.

London and pursued his studies until his death in 1723 at the age of 91. Leeuwenhoek refused to sell his microscopes.

After Leeuwenhoek's death, microbiology did not advance for more than a century. Eventually microscopes became more widely available, and progress resumed. Several workers discovered ways to stain microorganisms with dyes to make them more visible. The Swedish botanist Carolus Linnaeus developed a general classification system for all living organisms. The German botanist Matthias Schleiden and the German zoologist Theodor Schwann formulated the **cell theory,** which states that cells are the fundamental units of life and carry out all the basic functions of living things. Today this theory still applies to all cellular organisms, but not to viruses.

The Germ Theory of Disease

The **germ theory of disease** states that microorganisms (germs) can invade other organisms and cause disease. Although this is a simple idea and is generally accepted today, it was not widely accepted when formulated in the mid-nineteenth century. Many people believed that broth, left standing, turned cloudy because of something about the broth itself. Even after it was shown that microorganisms in the broth caused it to turn cloudy, people believed that the microorganisms, like the "worms" (fly larvae, or maggots) in rotting meat, arose from nonliving things, a concept known as **spontaneous generation**. Widespread belief in spontaneous generation, even among scientists, hampered further development of the science of microbiology and the acceptance of the germ theory of disease. As long as they believed that microorganisms could arise from nonliving substances, scientists saw no purpose in considering how diseases were transmitted or how they could be controlled. Dispelling the belief in spontaneous generation took years of painstaking effort.

The Roman writer, Varro, suggested in the first century. B.C. that people might get sick from inhaling "tiny animals."

Early Studies

For as long as humans have existed, some probably have believed that living things somehow originated spontaneously from nonliving matter. Aristotle's theories about his four "elements"—fire, earth, air, and water—seem to have suggested that nonliving forces somehow contributed to the generation of life. Even some naturalists believed that rodents arose from moist grain, beetles from dust, and worms and frogs from mud. As late as the nineteenth century, it seemed obvious to most people that rotting meat gave rise to "worms."

In the late seventeenth century, the Italian physician Francesco Redi devised a set of experiments to demonstrate that if pieces of meat were covered with gauze so that flies could not reach them, no "worms" appeared in the meat, no matter how rotten it was (➢ Figure 1.6). Maggots did, however, hatch from fly eggs laid on top of the gauze. Despite proof that maggots did not arise spontaneously, some scientists, such as the English clergyman John Needham, still believed in spontaneous generation—at least that of microorganisms. Lazzaro Spallanzani, an Italian cleric and scientist, was more skeptical. He boiled broth infusions containing organic (living or previously living) matter and sealed the flasks to demonstrate that no organisms would de-

➢ Figure 1.6 **Redi's experiments refuting the spontaneous generation of maggots in meat.** When meat is exposed in an open jar, flies lay their eggs on it, and the eggs hatch into maggots (fly larvae). In a sealed jar, however, no maggots appear. If the jar is covered with gauze, maggots hatch from eggs that the flies lay on top of the gauze, but still no maggots appear in the meat.

➢ **Figure 1.7 A "swan-necked" flask that Pasteur used in refuting the theory of spontaneous generation.** Although air could enter the flasks, microbes became trapped in the curved necks and never reached the contents. The contents, therefore, remained sterile—and still are—despite their exposure to the air.

velop spontaneously in them. Critics did not accept this as disproof of spontaneous generation. They argued that boiling drove off oxygen (which they thought all organisms required) and that sealing the flasks prevented its return.

Several scientists tried different ways of introducing air to counter this criticism. Schwann heated air before introducing it into flasks, and other scientists filtered air through chemicals or cotton plugs. All these methods prevented the growth of microorganisms in the flasks. But the critics still argued that altering the air prevented spontaneous generation.

Even nineteenth-century scientists of some stature continued to argue vociferously in favor of spontaneous generation. They believed that an organic compound previously formed by living organisms contained a "vital force" from which life sprang. The force, of course, required air, and they believed that all the methods of introducing air somehow changed it so that it could not interact with the force.

The proponents of spontaneous generation were finally defeated, mainly by the work of the French chemist Louis Pasteur and the English physicist John Tyndall. When the French Academy of Science sponsored a competition in 1859 "to try by well-performed experiment to throw new light on the question of spontaneous generation," Pasteur entered the competition.

During the years Pasteur worked in the wine industry, he had established that alcohol was produced in wine only if yeast was present, and he learned a lot about the growth of microorganisms. Pasteur's experiment for the competition involved his famous "swan-necked" flasks (➢ Figure 1.7). He boiled *infusions* (broths of foodstuffs) in flasks, heated the glass necks, and drew them out into long, curved tubes open at the end. Air could enter the flasks without being subjected to any of the treatments that critics had claimed destroyed its effectiveness. Airborne microorganisms could also enter the necks of the flasks, but they became trapped in the curves of the neck and never reached the infusions. The infusions from Pasteur's experiments remained sterile unless a flask was tipped so that the infusion flowed into the neck and back into the flask. This manipulation allowed microorganisms trapped in the neck to wash into the infusion, where they could grow and cause the infusion to become cloudy. In another experiment Pasteur filtered air through three cotton plugs. He then immersed the plugs in sterile infusions, demonstrating that growth occurred in the infusions from organisms trapped in the plugs.

Tyndall delivered another blow to the idea of spontaneous generation when he arranged sealed flasks of boiled infusion in an airtight box. After allowing time for all dust particles to settle to the bottom of the box, he carefully removed the covers from the flasks. These flasks, too, remained sterile. Tyndall had shown that air could be sterilized by settling, without any treatment that would prevent the "vital force" from acting.

Both Pasteur and Tyndall were fortunate that the organisms present in their infusions at the time of boiling were destroyed by heat. Others who tried the same experiments observed that the infusions became cloudy from growth of microorganisms. We now know that the infusions in which growth occurred contained heat-resistant or spore-forming organisms, but at the time, the growth of such organisms was seen as evidence of spontaneous generation. Still, the works of Pasteur and Tyndall successfully disproved spontaneous generation to most scientists of the time. Recognition that microbes must be introduced into a medium before their growth can be observed paved the way for further development of microbiology—especially for development of the germ theory of disease.

Pasteur's Further Contributions

Pasteur was not a physician and had never practiced medicine when he was asked to treat his first case of human exposure to rabies.

Louis Pasteur (➢ Figure 1.8) was such a giant among nineteenth-century scientists working in microbiology that we must consider some of his many contributions. Born in 1822, the son of a sergeant in Napoleon's army, Pasteur worked as a portrait painter and a teacher before he began to study chemistry in his spare time. Those studies led to posts in several French universities as professor of chemistry and to significant contributions to the wine and silkworm industries. He discovered that carefully selected yeasts made good wine, but that mixtures of other microorganisms competed with the yeast for sugar and made wine taste oily or sour. To combat this problem, Pasteur developed the technique of pasteurization (heating wine to 56°C in the absence of oxygen for 30 minutes) to kill unwanted organisms. While studying silkworms, he identified three different microorganisms, each of which caused a different disease.

➢ Figure 1.8 **Louis Pasteur in his laboratory.** The first rabies vaccine, developed by Pasteur, was made from the dried spinal cords of infected rabbits.

His association of specific organisms with particular diseases, even though in silkworms rather than in humans, was an important first step in proving the germ theory of disease.

Despite personal tragedy—the deaths of three daughters and a cerebral hemorrhage that left him with permanent paralysis—Pasteur went on to contribute to the development of vaccines. The best known of his vaccines is the rabies vaccine, made of dried spinal cord from rabbits infected with rabies, which was tested in animals. When a 9-year-old boy who had been severely bitten by a rabid dog was brought to him, Pasteur administered the vaccine. The boy, who had been doomed to die, survived and became the first person to be immunized against rabies.

In 1894 Pasteur became director of the Pasteur Institute, which was built for him in Paris. Until his death in 1895, he guided the training and work of other scientists at the Institute. Today the Pasteur Institute is a thriving research center—an appropriate memorial to its founder.

Robert Koch disproved the idea that tuberculosis was inherited.

Koch's Contributions

Robert Koch (➢ Figure 1.9), a contemporary of Pasteur's, finished his medical training in 1872 and worked as a physician in Germany throughout most of his career. After he bought a microscope and photographic equipment, he spent most of his time studying bacteria, especially those that cause disease. Koch identified the bacterium that causes anthrax, a highly contagious and lethal disease in cattle and sometimes in humans. He recognized both actively dividing cells and dormant cells (spores) and developed techniques for studying them *in vitro* (outside a living organism).

Koch also found a way to grow bacteria in *pure cultures*—cultures that contained only one kind of organism. He tried streaking bacterial suspensions on potato slices and then on solidified gelatin. But gelatin melts at incubator (body) temperature; even at room temperature, some microbes liquefy it. Finally, Angelina Hesse, the American wife of one of Pasteur's colleagues, suggested that Koch add agar (a thickener used in cooking) to his bacteriological media. This created a firm surface over which microorganisms could be spread very thinly—so thinly that some individual organisms were separated from all others. Each individual organism then multiplied to make a colony of thousands of descendants. Koch's technique of preparing pure cultures is still used today.

Koch's outstanding achievement was the formulation of four postulates to associate a particular organism with a specific disease. **Koch's postulates,** which provided scientists with a method of establishing the germ theory of disease, are as follows:

1. The specific causative agent must be found in every case of the disease.
2. The disease organism must be isolated in pure culture.
3. Inoculation of a sample of the culture into a healthy, susceptible animal must produce the same disease.
4. The disease organism must be recovered from the inoculated animal.

➢ Figure 1.9 **Robert Koch.** Koch formulated four postulates for linking a given organism to a specific disease.

Implied in Koch's postulates is his one organism–one disease concept. The postulates assume that an infectious disease is caused by a single organism, and they are directed toward estab-

What's in the Last Drop?

During the nineteenth century, French and German scientists were fiercely competitive. One area of competition was the preparation of pure cultures. Koch's reliable method of preparing pure cultures from colonies on solid media allowed German microbiologists to forge ahead. The French microbiologists' method of broth dilution, though now often used to count organisms (Chapter 6), hampered their progress. They added a few drops of a culture to fresh broth, mixed it, and added a few drops of the mixture to more fresh broth. After several successive dilutions, they assumed that the last broth that showed growth of microbes had contained a single organism. Unfortunately, the final dilution often contained more than one organism, and sometimes the organisms were of different kinds. This faulty technique led to various fiascos, such as inoculating animals with deadly organisms instead of vaccinating them.

lishing that fact. This concept also was an important advance in the development of the germ theory of disease.

After obtaining a laboratory post at Bonn University in 1880, Koch was able to devote his full time to studying microorganisms. He identified the bacterium that causes tuberculosis and developed a complex method of staining this organism. He also guided the research that led to the isolation of *Vibrio cholerae,* the bacterium that causes cholera.

In a few years Koch became professor of hygiene at the University of Berlin, where he taught a microbiology course believed to be the first ever offered. He also developed *tuberculin,* which he hoped would be a vaccine against tuberculosis. Because he underestimated the difficulty of killing the organism that causes tuberculosis, use of tuberculin resulted in several deaths from that disease. Although tuberculin was unacceptable as a vaccine, its development laid the groundwork for a skin test to diagnose tuberculosis. After the vaccine disaster, Koch left Germany. He made several visits to Africa, at least two visits to Asia, and one to the United States.

In the remaining 15 years of his life, his accomplishments were many and varied. He conducted research on malaria, typhoid fever, sleeping sickness, and several other diseases. His studies of tuberculosis won him the Nobel Prize for physiology or medicine in 1905, and his work in Africa and Asia won him great respect on those continents.

Work Toward Controlling Infections

Like Koch and Pasteur, two nineteenth-century physicians, Ignaz Philipp Semmelweis of Austria and Joseph Lister of England, were convinced that microorganisms caused infections (➢ Figure 1.10). Semmelweis recognized a connection between autopsies and puerperal (childbed) fever. Many physicians went directly from performing autopsies to examining women in labor without so much as washing their hands. When Semmelweis attempted to encourage more sanitary practices, he was ridiculed and harassed until he had a nervous breakdown and was sent to an asylum. Ultimately, he suffered the curious irony of succumbing to an infection caused by the same organism that produces puerperal fever. In 1865, Lister, who had read of Pasteur's work on pasteurization and Semmelweis's work on improving sanitation, initiated the use of dilute carbolic acid on bandages and instruments to reduce infection. Lister, too, was ridiculed, but with his imperturbable temperament, resolute will, and tolerance of hostile criticism, he was able to continue his work. His methods, the first *aseptic techniques,* were proven effective by the decrease in surgical wound infections in his surgical wards. At age 75, some 37 years after he introduced the use of carbolic acid, Lister was awarded the Order of Merit for his work in preventing the spread of infection. He is considered the father of antiseptic surgery.

- ✓ What similarities and differences do you see when comparing past epidemics of plague with today's AIDS epidemic?
- ✓ State the germ theory of disease. Try to think of an explanation of disease causation which would be contrary to the germ theory.
- ✓ How did Pasteur's "swan-necked" flasks experiment disprove the theory of spontaneous generation?
- ✓ Why was the French microbiologists' method of broth dilution inadequate for obtaining pure cultures of organisms?

Emergence of Special Fields of Microbiology

Pasteur, Koch, and most other microbiologists considered to this point were generalists interested in a wide variety of problems. Certain other contributors to microbiology had more specialized interests, but their achievements were no less valuable. In fact, those achievements helped establish the special fields of immunology, virology, chemotherapy, and microbial genetics—fields that are today prolific research areas. Selected fields of microbiology are defined in Table 1.2.

Immunology

Disease depends not only on microorganisms invading a host but also on the host's response to that invasion. Today, we know that the host's response is in part a response of the immune system.

The ancient Chinese knew that a person scarred by smallpox would not again get the disease. They took dried scabs from lesions of people who were recovering from the disease and ground them into a powder that they sniffed. As a result of inhaling weakened organisms, they acquired a mild case of smallpox but were protected against subsequent infection.

Pocahontas died of smallpox in 1617 in England.

Smallpox was unknown in Europe until the Crusaders carried it back from the Near East in the twelfth century. By the seventeenth century it was widespread. In 1717 Lady Ashley Montagu, wife of the British ambassador to Turkey, introduced a kind of immunization to England. A thread was soaked in fluid from a smallpox vesicle (blister) and drawn through a small incision in the arm. This technique, called *variolation,* was used at first by only a few prominent people, but eventually it became widespread.

In the late eighteenth century, Edward Jenner (➢ Figure 1.11) realized that milkmaids who got cowpox did not get smallpox, and he inoculated his own son with fluid from a cowpox blister. He later similarly inoculated an 8-year-old and subsequently inoculated the same child with smallpox. The child remained healthy. The word *vaccinia* (*vacca,* the Latin name for cow) gave rise both to the name of the virus that causes cowpox and to the word *vaccine*. In the early 1800s Jenner received grants amounting to a total of 30,000 British pounds to extend his work on vaccination. Today, those grants would be worth more than $1 million. They may have been the first grants for medical research.

Pasteur contributed significantly to the emergence of immunology with his work on vaccines for rabies and cholera. In 1879, when Pasteur was studying chicken cholera, his assistant accidentally used an old chicken cholera culture to inoculate some chickens. The chickens did not develop disease symptoms. When the assistant later inoculated the same chickens with a fresh chicken cholera culture, they remained healthy. Although he had not planned to use the old culture first, Pasteur did realize that the chickens had been immunized against chicken cholera. He reasoned that the organisms must have lost their ability to produce disease but retained their ability to produce immunity. This finding led Pasteur to look for techniques that would have the same effect on other organisms. His development of the rabies vaccine was a successful attempt.

Along with Jenner and Pasteur, the nineteenth-century Russian zoologist Elie Metchnikoff was a pioneer in immunology (➢ Figure 1.12). In the 1880s many scientists believed immunity was due to noncellular substances in the blood. Metchnikoff discovered that certain cells in the body could ingest microbes. He named those cells *phagocytes,* which literally means "cell-eating." The identification of phagocytes as cells that defend the body against invading microorganisms was a first step in understanding immunity. Metchnikoff also developed several vaccines. Some were successful, but unfortunately some infected the recipients with the microorganisms against which they were supposedly being immunized. A few of his subjects acquired gonorrhea and syphilis from his vaccines.

Virology

The science of virology emerged after that of bacteriology because viruses could not be recognized until certain techniques for studying and isolating larger particles such as bacteria had been developed. When Pasteur's collaborator Charles Chamberland developed a porcelain filter to remove bacteria from water in 1884, he had no idea that any kind of infectious agent could pass through the filter. But researchers soon realized that some filtrates (materials that passed through the filters) remained infectious even after the bacteria were filtered out. The Dutch microbiologist Martinus Beijerinck determined why such filtrates were infectious and

➢ Figure 1.11 **Edward Jenner vaccinating a child against smallpox.**

(a)

(b)

➢ Figure 1.10 **Two nineteenth-century pioneers in the control of infections.** (a) Ignaz Philipp Semmelweis, who died in an asylum before his innovations were widely accepted, depicted on a 1965 Austrian postage stamp; (b) Joseph Lister, who successfully carried on Semmelweis's work.

Calves, shaved, inoculated with cowpox, and covered with lesions, were led house to house by entrepreneurs offering vaccination during colonial times.

A "Thorny" Problem

Metchnikoff's personal life played a role in his discovery of phagocytes. Becoming despondent after the death of his first wife in 1873, he tried to kill himself with an overdose of opium. He did not due, but married again in 1875. The prenuptial agreement for his second marriage included taking his wife's dozen younger brothers and sisters into his home. On one occasion while Metchnikoff was at lunch, the young pranksters poked rose thorns into a starfish he was studying. Upon return, he found cells of the starfish gathered around the thorns. Study of these cells showed that they could devour foreign substances (phagocytosis)—an important body defense. When his second wife became very ill with typhoid fever in 1880, he again tried to take his own life. This time his method of suicide was an experiment—self-injection with relapsing fever to find out if the disease was transmissible by blood. He was awarded the Nobel Prize in 1908 for his work on immunology and phagocytosis.

➢ Figure 1.12 **Elie Metchnikoff.** Metchnikoff was one of the first scientists to study the body's defenses against invading microorganisms.

was thus the first to characterize viruses. The term *virus* had been used earlier to refer to poisons and to infectious agents in general. Beijerinck used the term to refer to specific *pathogenic* (disease-causing) molecules incorporated into cells. He also believed that these molecules could borrow for their own use existing metabolic and replicative mechanisms of the infected cells, known as *host cells*.

Further progress in virology required development of techniques for isolating, propagating, and analyzing viruses. The American scientist Wendell Stanley crystallized tobacco mosaic virus in 1935, showing that an agent with properties of a living organism also behaved as a chemical substance (➢ Figure 1.13). The crystals consisted of protein and ribonucleic acid (RNA). The nucleic acid was soon shown to be important in the infectivity of viruses. Viruses were first observed with an electron microscope in 1939. From that time both chemical and microscopic studies were used to investigate viruses.

By 1952 the American biologists Alfred Hershey and Martha Chase had demonstrated that the genetic material of some viruses is another nucleic acid, deoxyribonucleic acid (DNA). In 1953 the American postdoctoral student James Watson and the English biophysicist Francis

(a)

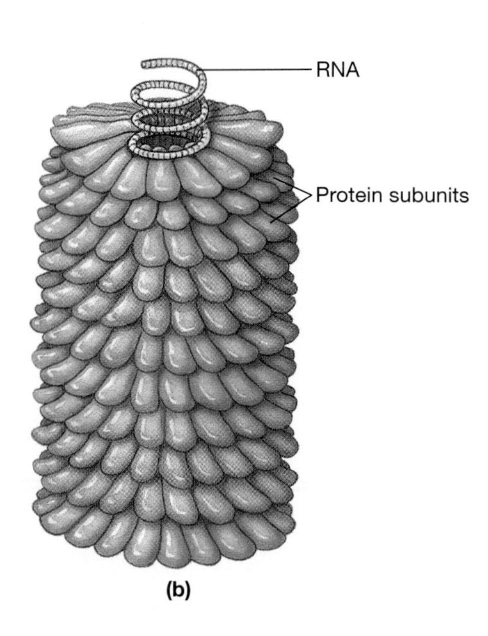

(b)

➢ Figure 1.13 **The tobacco mosaic virus.** (a) Electron micrograph of tobacco mosaic virus (magnification approx. 617,000X). (b) The structure of the tobacco mosaic virus. A helical core of RNA is surrounded by a coat that consists of repeating protein units. The structure of the particles is so regular that the viruses can be crystallized.

Crick determined the structure of DNA. The stage was set for rapid advances in understanding how DNA functions as genetic material both in viruses and in cellular organisms. Since the 1950s hundreds of viruses have been isolated and characterized. Although much remains to be learned about viruses, tremendous progress has been made in understanding their structure and how they function.

Chemotherapy

The Greek physician Dioscorides compiled *Materia Medica* in the first century A.D. This five-volume work listed a number of substances derived from medicinal plants still in use today—digitalis, curare, ephedrine, and morphine—along with a number of herbal medications. Credit for bringing herbal medicine to the United States is given to many groups of settlers, but Native Americans used many medicinal plants before the arrival of Europeans in the Americas. Many so-called primitive peoples still use herbs extensively, and some pharmaceutical companies finance expeditions into the Amazon Basin and other remote areas to investigate the uses the natives make of the plants around them.

During the Middle Ages virtually no advances were made in the use of chemical substances to treat diseases. Early in the sixteenth century the Swiss physician Aureolus Paracelsus used metallic chemical elements to treat diseases—antimony for general infections and mercury for syphilis. In the mid-seventeenth century Thomas Sydenham, an English physician, introduced cinchona tree bark to treat malaria. This bark, which we now know contains quinine, had been used to treat fevers in Spain and South America. In the nineteenth century morphine was extracted from the opium poppy and used medicinally to alleviate pain.

Paul Ehrlich, the first serious researcher in the field of chemotherapy (➢ Figure 1.14), received his doctoral degree from the University of Leipzig, Germany, in 1878. His discovery that certain dyes stained microorganisms but not animal cells suggested that the dyes or other chemicals might selectively kill microbial cells. This led him to search for the "magic bullet," a chemical that would destroy specific bacteria without damaging surrounding tissues. Ehrlich coined the term *chemotherapy* and headed the world's first institute concerned with the development of drugs to treat disease.

Early in the twentieth century the search for the magic bullet continued, especially among scientists at Ehrlich's institute. After testing hundreds of compounds (and numbering each compound), Ehrlich found compound 418 (arsenophenylglycine) to be effective against sleeping sickness and compound 606 (Salvarsan) to be effective against syphilis. For 40 years Salvarsan remained the best available treatment for this disease. In 1922 Alexander Fleming, a Scottish

Informed Consent

During the Panama Canal construction in 1905, yellow fever struck the workers as they struggled in the swamps. Yellow fever was an often fatal disease. Because the entire canal project was jeopardized, the American physician, Walter Reed, was asked to control the disease. Reed listened to Cuban physician, Carlos Finlay y Barres, who claimed that yellow fever was carried by mosquitoes, and ignored those who insisted that yellow fever was due to swamp air.

Reed designed experiments to determine the role of mosquitoes or swamp air in the disease's transmission. Volunteers were bitten by mosquitoes known to have bitten yellow fever patients, were injected with blood from patients, or slept in sealed huts on sheets soaked with patients' vomitus. Many of those bitten by mosquitoes or injected with patient's blood died from yellow fever. Those who slept on the vomit-soaked sheets, wore the clothing of the dead, and ate from their dishes did not become ill. It is not known how Reed got volunteers for such a deadly experiment. Why do you think these men volunteered for this deadly task?

➢ Figure 1.14 **Paul Ehrlich.** Ehrlich was a pioneer in the development of chemotherapy for infectious disease.

physician, discovered that lysozyme, an enzyme found in tears, saliva, and sweat, could kill bacteria. Lysozyme was the first body secretion shown to have chemotherapeutic properties.

The development of antibiotics began in 1917 with the observation that certain bacteria (actinomycetes) stopped the growth of other bacteria. In 1928 Fleming (➢ Figure 1.15) observed that a colony of *Penicillium* mold contaminating a culture of *Staphylococcus* bacteria had prevented growth of bacteria adjacent to itself. Although he was not the first to observe this phenomenon, Fleming did recognize its potential for countering infections. However, purification of sufficient quantities of the substance he called *penicillin* proved to be very difficult. The great need for such a drug during World War II, money from the Rockefeller Institute, and the hard work of the German biochemist Ernst Chain, the Australian pathologist Howard Florey, and researchers at Oxford University accomplished the task. Penicillin became available as a safe and versatile chemotherapeutic agent for use in humans.

While this work was going on, sulfa drugs were being developed. In 1935, prontosil rubrum, a reddish dye containing a sulfonamide chemical group, was used in treating streptococcal infections. Further study showed that sulfonamides were converted in the body to sulfanilamides; much subsequent work was devoted to developing drugs containing sulfanilamide. The German chemist Gerhard Domagk played an important role in this work, and one of the drugs, prontosil, saved the life of his daughter. In 1939 he was awarded a Nobel Prize for his work, but Hitler refused to allow him to make the trip to receive it. Extensions of Domagk's work led to the development of isoniazid, an effective agent against tuberculosis. Both sulfa drugs and isoniazid are still used today.

The development of antibiotics resumed with the work of Selman Waksman, who was born in Ukraine and moved to the United States in 1910. Inspired by the 1939 discovery, by the French microbiologist Rene Dubos of tyrothricin, an antibiotic produced by soil bacteria, Waksman examined soil samples from all over the world for growth-inhibiting microorganisms or their products. He coined the term *antibiotic* in 1941 to describe actinomycin and other products he isolated. Both tyrothricin and actinomycin proved to be too toxic for general use as antibiotics. After repeated efforts, Waksman isolated the less toxic drug streptomycin in 1943. Streptomycin constituted a major breakthrough in the treatment of tuberculosis. In the same decade Waksman and others isolated neomycin, chloramphenicol, and chlortetracycline.

Examining soil samples proved to be a good way to find antibiotics, and explorers and scientists still collect soil samples for analysis. The more common antibiotic-producing organisms are rediscovered repeatedly, but the possibility of finding a new one always remains. Even the sea has yielded antibiotics, especially from the fungus *Cephalosporium acremonium*. The Italian microbiologist Giuseppe Brotzu noted the absence of disease organisms in seawater where sewage entered, and he determined that an antibiotic must be present. Cephalosporin was subsequently purified, and a variety of cephalosporin derivatives are now available for treating human diseases.

➢ Figure 1.15 **Alexander Fleming.** Fleming discovered the antibacterial properties of penicillin.

How Microbiologists Investigate Problems

Like other scientists, microbiologists investigate problems by designing and carrying out experiments. Such experiments have provided the information health scientists apply to solve medical problems. Much of this text is devoted to presenting information obtained from experiments and to showing how that information is used in understanding infectious diseases. However, we believe that all science students will be interested in knowing how scientific problems are investigated.

First, a scientific problem must concern some aspect of the natural world because scientific methods can deal only with natural conditions and events. Microbiological problems deal with natural conditions and events involving microbes. Second, scientific problems must be clearly defined and sufficiently limited in scope so that a hypothesis and a prediction can be formulated. A **hypothesis** is a tentative explanation to account for an observed condition or event. The hypothesis in a particular experiment must be (1) an explanation for the defined problem and (2) testable. A testable hypothesis is one for which evidence can be collected to support or refute the hypothesis. A **prediction** is an outcome or consequence that will result if the hypothesis is true. Before beginning a scientific experiment, one must define the problem and make a hypothesis and a prediction.

A good hypothesis is one that offers the most reasonable explanation and the simplest solution to a problem. The purpose of scientific experiments is to test hypotheses by determining the correctness of predictions derived from the hypotheses. Scientific progress is made by making and testing hypotheses.

For example, suppose that a microbiologist has isolated an organism in pure culture and wants to know the effects of temperature on its growth. On the basis of information from prior research, he or she might (1) hypothesize that the organism's growth rate increases with temperature and (2) predict that the rate of increase in the number of organisms in a culture is proportional to the increase in temperature. After making the hypothesis and a prediction, the investigator designs an experiment to test the hypothesis. The experiment must be designed specifically to test the hypothesis and to collect evidence to determine whether the prediction is true.

To design a good experiment, an investigator must consider all variables that might affect the outcome. A **variable** is anything that can change for the purposes of an experiment. An experiment should have only one **experimental variable,** the factor that is purposely changed for the experiment. For example, in the study of the effects of temperature on the growth of an organism, temperature is the experimental variable. The hypothesis and prediction are related to the experimental variable. All other variables are **control variables,** factors that *can* change but that are prevented from changing for the duration of the experiment. In our example, the control variables include the number and characteristics of the organism, the quantity and properties of the medium, and all environmental factors except temperature.

When all variables have been identified, the investigator establishes the procedures for carrying out the experiment. Once the experiment has been designed, it must be carried out exactly as planned. All observations must be made and recorded accurately and precisely. If problems or unusual situations are encountered, they must be noted carefully. For example, should an incubator fail to maintain certain cultures at the proper temperature for the appropriate length of time, this failure should be noted and taken into consideration in interpreting the experiment. When the experiment is completed, the researcher analyzes and interprets the results in light of the hypothesis and prediction. The analysis of the results of an experiment often involves preparation of tables and graphs and usually compares results obtained under experimental and control conditions.

The goal of an experiment is to draw conclusions as to whether the prediction is true. If the experimental results, when analyzed, do not support the hypothesis, they may nevertheless suggest a better alternative hypothesis. The experimenter might wish to design further experiments to test this new hypothesis. Often, it is unexpected experimental results that lead to the most interesting discoveries.

In this textbook, you will find some boxes, such as A Winter Dilemma (p. 31) and What Grows in Your Health and Beauty Aids? (p. 554) that suggest projects that you might try yourselves. Even if you are not able to try these projects, at least make a mental plan of what hypotheses and steps you could use to investigate such a problem. Perhaps you can discuss your experimental designs during lecture or in lab. What other questions occur to you? How would you go about forming and testing hypotheses for them? In addition to performing experiments, scientists also should report their results so that other scientists can verify and use the information. Scientific knowledge increases by the sharing of information. This allows other scientists to repeat experiments and determine whether the results are reproducible. It also allows them to develop new experiments that build on existing information.

The fact that many antibiotics have been discovered does not stop the search for more. As long as there are untreatable infectious diseases, the search will continue. Even when effective treatment becomes available, it is always possible that a better, less toxic, or cheaper treatment can be found. Of the many chemotherapeutic agents currently available, none can cure viral infections. Consequently, much of today's drug research is focused on developing effective antiviral agents.

Genetics and Molecular Biology

Modern genetics began with the rediscovery in 1900 of Gregor Mendel's principles of genetics. Even after this significant event, for nearly three decades little progress was made in understanding how microbial characteristics are inherited. For this reason, microbial genetics is the youngest branch of microbiology. In 1928 the British scientist Frederick Griffith discovered that previously harmless bacteria could change their own nature so as to become capable of causing disease. The remarkable thing about this discovery was that live bacteria were shown to acquire heritable traits from dead ones. During the early 1940s, Oswald Avery, Maclyn McCarty, and Colin MacLeod of the Rockefeller Institute in New York City demonstrated that the change was produced by DNA. After that finding came the crucial discovery of the structure of DNA by James Watson and Francis Crick. This breakthrough ushered in the modern era of molecular genetics.

About the same time, the American geneticists Edward Tatum and George Beadle used genetic variations in the mold *Neurospora* to demonstrate how genetic information controls metabolism. In the early 1950s the American geneticist Barbara McClintock discovered that some genes (units of inherited information) can move from one location to another on a chromosome. Before McClintock's work, genes were thought to remain stationary. Her revolutionary discovery forced geneticists to revise their thinking about genes.

More recently, scientists have discovered the genetic basis that underlies the human body's ability to make an enormous diversity of *antibodies,* molecules that the immune system produces to combat invading microbes and their toxic products. Within cells of the immune system, genes are shuffled about and spliced together in various combinations, allowing the body to make millions of different antibodies, including some that can protect us from threats that the body has never previously encountered.

Tomorrow's History

Today's discovery is tomorrow's history. In an active research field such as microbiology, it is impossible to present a complete history. Some of the microbiologists omitted from this discussion are listed in Table 1.3. The period represented there, 1874–1917, is called the Golden Age of Microbiology. Many terms used to describe these scientists' accomplishments will be unfamiliar, but you will become familiar with them as you pursue the study of microbiology. Since 1900, Nobel Prizes have been awarded annually to outstanding scientists, many of whom were in the fields of physiology or medicine (Table 1.4). In some years the prize has been shared by several scientists, although the scientists may have made independent contributions. Refer to Tables 1.3 and 1.4 as you begin to study each new area of microbiology.

You can see from Table 1.4 that microbiology has been in the forefront of research in medicine and biology for several decades, and probably never more so than today. One reason is the renewed focus on infectious disease brought about by the advent of AIDS. Another is the dramatic progress in genetic engineering that has been made in the past two decades. Microorganisms have been and continue to be an essential part of the genetic engineering revolution. Most of the key discoveries that led to our present understanding of genetics emerged from research with microbes. Today scientists are attempting to redesign microorganisms for a variety of purposes (as we will see in Chapter 8). Bacteria have been converted into factories that produce drugs, hormones, vaccines, and a variety of biologically important compounds. And microbes, viruses in particular, are often the vehicle by which scientists insert new genes into

➢ Table 1.3 **The Golden Age of microbiology: Early microbiologists and their achievements**

Year	Investigator	Achievement
1874	Billroth	Discovery of round bacteria in chains
1876	Koch	Identification of *Bacillus anthracis* as causative agent of anthrax
1878	Koch	Differentiation of staphylococci
1879	Hansen	Discovery of *Mycobacterium leprae* as causative agent of leprosy
1880	Neisser	Discovery of *Neisseria gonorrhoeae* as causative agent of gonorrhea
1880	Laveran and Ross	Identification of life cycle of malarial parasites in red blood cells of infected humans
1880	Eberth	Discovery of *Salmonella typhi* as causative agent of typhoid fever
1880	Pasteur and Sternberg	Isolation and culturing of pneumonia cocci from saliva
1881	Koch	Animal immunization with attenuated anthrax bacilli
1882	Leistikow and Loeffler	Cultivation of *Neisseria gonorrhoeae*
1882	Koch	Discovery of *Mycobacterium tuberculosis* as causative agent of tuberculosis
1882	Loeffler and Schutz	Identification of actinobacillus that causes the animal disease glanders
1883	Koch	Identification of *Vibrio cholerae* as causative agent of cholera
1883	Klebs	Identification of *Corynebacterium diphtheriae* and toxin as causative agent of diphtheria
1884	Loeffler	Culturing of *Corynebacterium diphtheriae*
1884	Rosenbach	Pure culturing of streptococci and staphylococci
1885	Escherich	Identification of *Escherichia coli* as a natural inhabitant of the human gut
1885	Bumm	Pure culturing of *Neisseria gonorrhoeae*
1886	Flugge	Staining to differentiate bacteria
1886	Fraenckel	*Streptococcus pneumoniae* related to pneumonia
1887	Weichselbaum	*Neisseria meningitidis* related to meningitis
1887	Bruce	Identification of *Brucella melitensis* as causative agent of brucellosis in cattle
1887	Petri	Invention of culture dish
1888	Roux and Yersin	Discovery of action of diphtheria toxin
1889	Charrin and Roger	Discovery of agglutination of bacteria in immune serum
1889	Kitasato	Discovery that *Clostridium tetani* produces tetanus toxin
1890	Pfeiffer	Identification of Pfeiffer bacillus, *Haemophilus influenzae*
1890	von Behring and Kitasato	Immunization of animals with diphtheria toxin
1892	Ivanovski	Discovery of filterability of tobacco mosaic virus
1894	Roux and Kitasato	Identification of *Yersinia pestis* as causative agent of bubonic plague
1894	Pfeiffer	Discovery of bacteriolysis in immune serum
1895	Bordet	Discovery of alexin (complement) and hemolysis
1896	Widal and Grunbaum	Development of diagnostic test based on agglutination of typhoid bacilli by immune serum
1897	van Ermengem	Discovery of *Clostridium botulinum* as causative agent of botulism
1897	Kraus	Discovery of precipitins
1897	Ehrlich	Formulation of side-chain theory of antibody formation
1898	Shiga	Discovery of *Shigella dysenteriae* as causative agent of dysentery
1898	Loeffler and Frosch	Discovery of filterability of virus that causes foot-and-mouth disease
1899	Beijerinck	Discovery of intracellular reproduction of tobacco mosaic virus
1901	Bordet and Gengou	Identification of *Bordetella pertussis* as causative agent of whooping cough; development of complement fixation test
1901	Reed and colleagues	Identification of virus that causes yellow fever
1902	Portier and Richet	Work on anaphylaxis
1903	Remlinger and Riffat-Bey	Identification of virus that causes rabies
1905	Schaudinn and Hoffmann	Identification of *Treponema pallidum* as causative agent of syphilis
1906	Wasserman, Neisser, and Bruck	Development of Wasserman reaction for syphilis antibodies
1907	Asburn and Craig	Identification of virus that causes dengue fever
1909	Flexner and Lewis	Identification of virus that causes poliomyelitis
1915	Twort	Discovery of viruses that infect bacteria
1917	d'Herelle	Independent rediscovery of viruses that infect bacteria (bacteriophages)

➢ Table 1.4 **Nobel prize awards for research involving microbiology**

Year of prize	Prize winner	Topic studied
1901	von Behring	Serum therapy against diphtheria
1902	Ross	Malaria
1905	Koch	Tuberculosis
1907	Laveran	Protozoa and the generation of disease
1908	Ehrlich and Metchnikoff	Immunity
1913	Richet	Anaphylaxis
1919	Bordet	Immunity
1928	Nicolle	Typhus exanthematicus
1939	Domagk	Antibacterial effect of prontosil
1945	Fleming, Chain, and Florey	Penicillin
1951	Theiler	Vaccine for yellow fever
1952	Waksman	Streptomycin
1954	Enders, Weller, and Robbins	Cultivation of polio virus
1958	Lederberg	Genetic mechanisms
	Beadle and Tatum	Transmission of hereditary characteristics
1959	Ochoa and Kornberg	Chemical substances in chromosomes that play a role in heredity
1960	Burnet and Medawar	Acquired immunological tolerance
1962	Watson, Crick, and Wilkins	Structure of deoxyribonucleic acid
1965	Jacob, Lwoff, and Monod	Regulatory mechanisms in microbial genes
1966	Rous	Viruses and cancer
1968	Holley, Khorana, and Nirenberg	Genetic code
1969	Delbruck, Hershey, and Luria	Mechanism of virus infection in living cells
1972	Edelman and Porter	Structure and chemical nature of antibodies
1975	Baltimore, Temin, and Dulbecco	Interactions between tumor viruses and genetic material of the cell
1976	Blumberg and Gajdusek	New mechanisms for the origin and dissemination of infectious diseases
1978	Smith, Nathans, and Arber	Restriction enzymes for cutting DNA
1980	Benacerraf, Snell, and Dausset	Immunological factors in organ transplants
1980	Berg	Recombinant DNA
1984	Milstein, Köhler, and Jerne	Immunology
1987	Tonegawa	Genetics of antibody diversity
1988	Black, Elion, and Hitchings	Principles of drug therapy
1989	Bishop and Varmus	Genetic basis of cancer
1990	Murray, Thomas, and Corey	Transplant techniques and drugs
1993	Mullis	Polymerase chain reaction method to amplify (copy) DNA
1993	Smith	Method to splice foreign components into DNA
1993	Sharp and Roberts	Genes can be discontinuou

other organisms. Such techniques are beginning to enable us to produce improved varieties of plants and animals such as pest-resistant crops (➢ Figure 1.16) and may even enable us to correct genetic defects in human beings.

In September 1990, a 4-year-old girl became the first gene-therapy patient. She had inherited a defective gene that crippled her immune system. Doctors at the National Institutes of Health (NIH) inserted a normal copy of the gene into some of her white blood cells in the laboratory and then injected these gene-treated cells back into her body, where, it was hoped, they will restore her immune system. Critics were worried that a new gene randomly inserted into her white blood cells could damage other genes and cause cancer. The experiment was a success and she enjoys good health today.

New information is constantly being discovered and sometimes supersedes earlier findings. Occasionally, new discoveries lead almost immediately to the development of medical appli-

➢ Figure 1.16 **Plant scientists using knowledge gained from research.** Microorganisms can be used to produce superior agricultural crops, such as strains that resist insect pests or that have greater productivity due to soil microorganisms that help the plants grow better. The poinsettia on the right has been inoculated with a combination of fungi and bacteria that help the plant grow better.

cations, as occurred with penicillin and as will most certainly occur when a cure or vaccine for AIDS is discovered. However, old ideas such as spontaneous generation and old practices such as unsanitary measures in medicine can take years to replace. Many new bioethics problems will require considerable thought. Decisions regarding AIDS testing and reporting, transplants, environmental cleanup, and related issues will not come easily or quickly. Because of the wealth of prior knowledge, it is likely that you will learn more about microbiology in a single course than many pioneers learned in a lifetime. Yet, those pioneers deserve great credit because they worked with the unknown and had few people to teach them.

Human Genome Project

Microbial genetic techniques have made possible the undertaking of a colossal and controversial scientific plan, the Human Genome Project. At a cost of approximately $3 billion over a period of about 15 years, this project is making progress in identifying the location and chemical sequence of all the genes in the human genome—that is, all the genetic material in the human species—now estimated to consist of about 3 billion base pairs. The project is expected to be completed by the year 2005. When finished, it will be like having the "owner's manual" for humankind. It will make possible an incredible array of manipulations of human genetics and functions. Researchers are now locating and sequencing genes via simple microbes at first, such as *Escherichia coli,* a common organism found in the colon and feces. They have completely sequenced *Haemophilus influenzae,* a cause of ear infections; *Helicobacter pylori,* the cause of stomach ulcers; *Borrelia burgdorferi,* which causes Lyme disease; plus several other bacteria, viruses, and a yeast. Knowing the genetics of pathogens will help us discover means of combating their subterfuges for evading our immune systems. And knowing about good microbes will help us to make them even more useful such as in cleaning up toxic and nuclear waste sites, manufacturing fertilizer, and producing new drugs. Within a few years we'll probably have the complete sequences for 50 to 100 microbes. Now investigators are shifting to human genes as techniques have become more efficient.

In 1991, less than 2000 human genes were known. Then Craig Ventner and his colleagues devised more efficient techniques, and in a few months doubled that number. Today there is some controversy over who should finish sequencing the human genome, and whether patenting of these genes should be allowed (e.g., by pharmaceutical houses). In any case, the project will doubtless finish ahead of its original deadline of the year 2005. Says Ventner, "The secrets of life are all spelled out for us in the genome, we just have to learn how to read it."

✓ What were the scientific contributions of: Jenner, Metchnikoff, Ehrlich, Fleming, McClintock, and Ventner?

✓ When was the Golden Age of Microbiology? What types of discoveries were mostly made during this period?

✓ What is the Human Genome Project? How has microbiology been associated with it?

Retracing Our Steps

Why Study Microbiology?

- Microorganisms are part of the human environment and are therefore important to human health and activities.
- The study of microorganisms provides insight into life processes in all forms of life.

Scope of Microbiology

The Microbes

- **Microbiology** is the study of all **microorganisms (microbes)** in the microscopic range. These include **bacteria, algae, fungi, viruses,** and **protozoa**.

The Microbiologists

- Immunology, virology, chemotherapy, and genetics are especially active research fields of microbiology.
- Microbiologists work as researchers or teachers in university, clinical, and industrial settings. They do basic research in the biological sciences; help to perform or devise diagnostic tests; develop and test antibiotics and vaccines; work to control infection, protect public health, and safeguard the environment; and play important roles in the food and beverage industries.

Historical Roots

- The ancient Greeks, Romans, and Jews all contributed to early understandings of the spread of disease.
- Diseases such as bubonic plague and syphilis caused millions of deaths because of the lack of understanding of how to control or treat the infections.
- The development of high-quality lenses by Leeuwenhoek made it possible to observe microorganisms and later to formulate the **cell theory**.

The Germ Theory of Disease

- The **germ theory of disease** states that microorganisms (germs) can invade other organisms and cause disease.

Early Studies

- Progress in microbiology and acceptance of the germ theory of disease required that the idea of **spontaneous generation** be refuted. Redi and Spallanzani demonstrated that organisms did not arise from nonliving material. Pasteur, with his swan-necked flasks, and Tyndall, with his dust-free air, finally dispelled the idea of spontaneous generation.

Pasteur's Further Contributions

- Pasteur also studied wine making and disease in silkworms and developed the first rabies vaccine. His association of particular microbes with certain diseases furthered the establishment of the germ theory.

Koch's Contributions

- Koch developed four postulates that aided in the definitive establishment of the germ theory of disease. **Koch's postulates** are as follows:
  1. The specific causative agent must be found in every case of the disease.
  2. The disease organism must be isolated in pure culture.
  3. Inoculation of a sample of the culture into a healthy, susceptible animal must produce the same disease.
  4. The disease organism must be recovered from the inoculated animal.
- Koch also developed techniques for isolating organisms, identified the bacillus that causes tuberculosis, developed tuberculin, and studied various diseases in Africa and Asia.

Work Toward Controlling Infections

- Lister and Semmelweis contributed to improved sanitation in medicine by applying the germ theory and using aseptic technique.

Emergence of Special Fields of Microbiology

Immunology

- Immunization was first used against smallpox; Jenner used fluid from cowpox blisters to immunize against it.
- Pasteur developed techniques to weaken organisms so they would produce immunity without producing disease.

Virology

- Beijerinck characterized viruses as pathogenic molecules that could take over a host cell's mechanisms for their own use.
- Reed demonstrated that mosquitoes carry the yellow fever agent, and several other investigators identified viruses in the early twentieth century. The structure of DNA, the genetic material in many viruses and in all cellular organisms, was discovered by Watson and Crick.
- New techniques for isolating, propagating, and analyzing viruses were developed. Viruses could then be observed and in many cases crystallized, and their nucleic acids could be studied.

Chemotherapy

- Substances derived from medicinal plants were virtually the only source of chemotherapeutic agents until Ehrlich began a systematic search for chemically defined substances that would kill bacteria.
- Fleming and his colleagues developed penicillin, and Domagk and others developed sulfa drugs.
- Waksman and others developed streptomycin and other antibiotics derived from soil organisms.

Genetics and Molecular Biology

- Griffith discovered that previously harmless bacteria could change their nature so as to become capable of causing disease. This genetic change was shown by Avery, McCarty, and MacLeod to be due to DNA. Tatum and Beadle studied biochemical mutants of *Neurospora* to show how genetic information controls metabolism.

Tomorrow's History

- Microbiology has been at the forefront of research in medicine and biology, and microorganisms continue to play a critical role in genetic engineering and gene therapy.

Human Genome Project

- The Human Genome Project will identify the location and sequence of all bases in the human genome. Microbes and microbiological techniques have contributed to this work.

Terminology Check

algae (*p. 3*)
antibiotics (*p. 2*)
bacteria (*p. 3*)
cell theory (*p. 11*)
control variable (*p. 19*)
experimental variable (*p. 19*)
fungi (*p. 4*)
germ theory of disease (*p. 11*)
hypothesis (*p. 19*)
Koch's postulates (*p. 13*)
microbe (*p. 3*)
microbiology (*p. 3*)
microorganism (*p. 1*)
prediction (*p. 19*)
protozoa (*p. 4*)
spontaneous generation (*p. 11*)
variable (*p. 19*)
viruses (*p. 4*)

See you on the Web

If you feel like you've got the lay of the land in this chapter, go check out the web for futher adventures. But before you do, make sure you are up to the challenge by taking at least some of these steps:

- Go back to the beginning of the chapter and make sure you can answer all of the questions I posed. Better yet, explain your answers to a classmate, roommate, friend, parent, or stranger on the bus. You'll be amazed at how your understanding of the material will be enhanced by explaining it to someone else.
- Review the concept check questions in the chapter. Answer them again to ensure that you have retained your understanding.
- Photocopy the key illustrations from this chapter, and use them to make your own flashcards. I think you'll find that actually making the flashcards is at least as beneficial as using them for review.
- Work through this same chapter in the companion Study Guide.

Once you are on the web, you will get one more chance to make sure you are on top of everything by taking a quiz for this chapter. It will challenge you with many different types of questions and then score your responses.

There are many other exciting features on the website, guaranteed to excite the explorer in you. To check them out, meet me at

http://www.wiley.com/college/black

Fundamentals of Chemistry

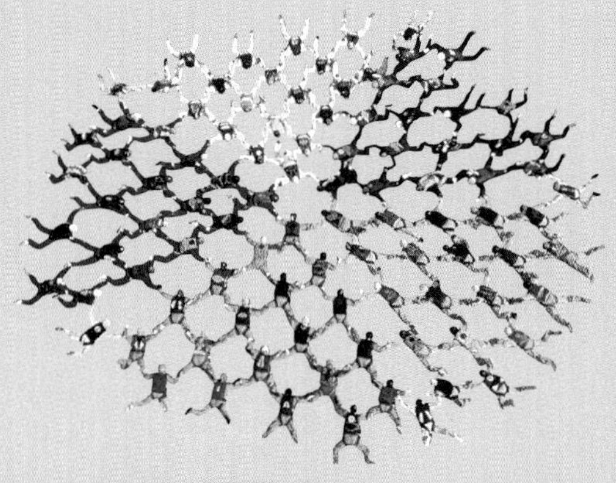

Free-falling and weightless, you climb toward the others. You link hands with your skydiving partners to form the pattern you discussed just moments ago in the safety of the airplane. Suspended between heaven and earth you consider the significance of the connections that bind you.

Just as the hands of these skydivers hold them in a complex pattern, chemical bonds between atoms hold together complex molecular patterns. The shapes of molecules are very important—when shapes change, the properties of the molecules themselves change. In an organism, this change can mean the difference between life and death.

All living and nonliving things, including microbes, are composed of matter. Thus, it is not surprising that all properties of microorganisms are determined by the properties of matter.

Why Study Chemistry?

Chemistry is the science that deals with the basic properties of matter. Therefore we need to know some chemistry to begin to understand microorganisms. Chemical substances undergo changes and interact with one another in *chemical reactions*. Metabolism, the use of nutrients for energy or for making the substance of cells, consists of many different chemical reactions. This is true regardless of whether the organism is a human or a microorganism. Thus, understanding the basic principles of chemistry is essential to understanding metabolic processes in living things. A microbiologist uses chemistry to understand the structure and function of microorganisms themselves and to understand how they affect humans in disease processes as well as how they affect all life on earth.

Chemical Building Blocks and Chemical Bonds

Chemical Building Blocks

Matter is composed of very small particles that form the basic chemical building blocks. Over the years, chemists have observed matter and deduced the characteristics of these particles. Just as the alphabet can be used to make thousands of words, the chemical building blocks can be used to make thousands of different substances. The complexity of chemical substances greatly exceeds the complexity of words. Words rarely contain more than 20 letters, whereas some complex chemical substances contain as many as 20,000 building blocks!

The smallest chemical unit of matter is the **atom.** Many different kinds of atoms exist. Matter composed of one kind of atom is called an **element.** Each element has specific properties that distinguish it from other elements. Carbon is an element; a pure sample of carbon consists of a vast number of carbon atoms. Oxygen and nitrogen also are elements; they are found as gases in the earth's atmosphere. Chemists use one- or two-letter symbols to designate elements—such as C for carbon, O for oxygen, N for nitrogen, and Na for sodium (from its Latin name, *natrium*).

Atoms combine chemically in various ways. Sometimes atoms of a single element combine with each other. For example, carbon atoms form long chains that are im-

Questions We'll Explore

A Why is knowledge of basic chemistry necessary to the understanding of microbiology?

B What terms describe the organization of matter, and which elements are found in living organisms?

C What are the properties of chemical bonds and chemical reactions?

D Which properties of water, solutions, colloidal dispersions, acids, and bases make them important in living things?

E What is organic chemistry, and what are the major functional groups of organic molecules?

F How do the structures and properties of carbohydrates contribute to their roles in living things?

G How do the structures and properties of fats, phospholipids, and steroids contribute to their roles in living things?

H How do the structures and properties of proteins, including enzymes, contribute to their roles in living things?

I How do the structures and properties of nucleotides contribute to their roles in living things?

portant in the structure of living things. Both oxygen and nitrogen form paired atoms, O_2 and N_2. More often, atoms of one element combine with atoms of other elements. Carbon dioxide (CO_2) contains one atom of carbon and two atoms of oxygen; water (H_2O) contains two atoms of hydrogen and one atom of oxygen. (The subscripts in these formulas indicate how many atoms of each element are present.)

When two or more atoms combine chemically, they form a **molecule.** Molecules can consist of atoms of the same element, such as N_2, or atoms of different elements, such as CO_2. Molecules made up of atoms of two or more elements are called **compounds.** Thus, CO_2 is a compound, but N_2 is not. The properties of compounds are different from those of their component elements. For example, in their elemental state, both hydrogen and oxygen are gases at ordinary temperatures. They can combine to form water, however, which is a liquid at ordinary temperatures.

Living things consist of atoms of relatively few elements, principally carbon, hydrogen, oxygen, and nitrogen, but these are combined into highly complex compounds. A simple sugar molecule, $C_6H_{12}O_6$, contains 24 atoms. Many molecules found in living organisms contain thousands of atoms.

The Structure of Atoms

Although the atom is the smallest unit of any element that retains the properties of that element, atoms do contain even smaller particles that together account for those properties. Physicists study many such subatomic particles, but we discuss only **protons, neutrons,** and **electrons.** Three important properties of these particles are atomic mass, electrical charge, and location in the atom (Table 2.1). *Atomic mass* is measured in terms of *atomic mass units (AMU).* The mass of a proton or a neutron is almost exactly equal to 1 AMU; electrons have a much smaller mass. With respect to electrical charge, electrons are negatively (−) charged, and protons are positively (+) charged. Neutrons are neutral, with no charge. Atoms normally have an equal number of protons and electrons and so are electrically neutral. The heavy protons and neutrons are densely packed into the tiny, central *nucleus* of the atom, whereas the lighter electrons move around the nucleus in what have commonly been described as orbits.

The atoms of a particular element always have the same number of protons; that number of protons is the **atomic number** of the element. Atomic numbers range from 1 to over 100. The numbers of neutrons and electrons in the atoms of many elements can change, but the number of protons—and therefore the atomic number—remains the same for all atoms of a given element.

Protons and electrons are oppositely charged. Consequently, they attract each other. This attraction keeps the electrons near the nucleus of an atom. The electrons are in constant, rapid motion, forming an electron cloud around the nucleus. Because some electrons have more energy than others, chemists use a model with concentric circles, or *electron shells,* to suggest different energy levels. Electrons with the least energy are located nearest the nucleus, and those with more energy are farther from the nucleus. Each energy level corresponds to an electron shell (➢ Figure 2.1).

An atom of hydrogen has only one electron, which is located in the innermost shell. An atom of helium has two electrons in that shell; two is the maximum number of electrons that can be found in the innermost shell. Atoms with more than two electrons always have two electrons in the inner shell and up to eight additional electrons in the second shell. The inner shell is filled before electrons are found in the second shell; the second shell is filled before electrons are found in the third shell, and so on. Very large atoms have several more electron shells of larger capacity, but in elements found in living things, the outer shell is chemically stable if

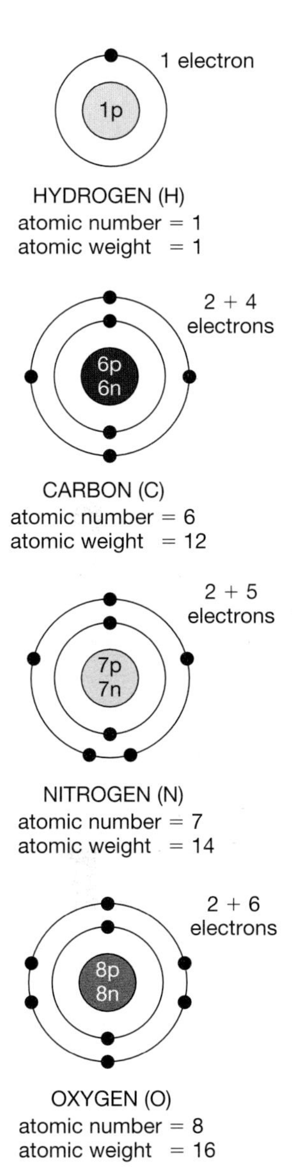

➢ Figure 2.1 **The structure of four biologically important atoms.** Hydrogen, the simplest element, has a nucleus consisting of a single proton and a single electron in the first shell. In carbon, nitrogen, and oxygen the first shell is filled with two electrons and the second shell is partly filled. Carbon, with six protons in its nucleus, has six electrons, four of them in the second shell. Nitrogen has five electrons, and oxygen six electrons in the second shell. It is the electrons in the outermost shell that take part in chemical bonding.

➢ Table 2.1 **Properties of atomic particles**

Particle	Atomic mass	Electrical charge	Location
Proton	1	+	Nucleus
Neutron	1	None	Nucleus
Electron	1/1836	−	Orbiting the nucleus

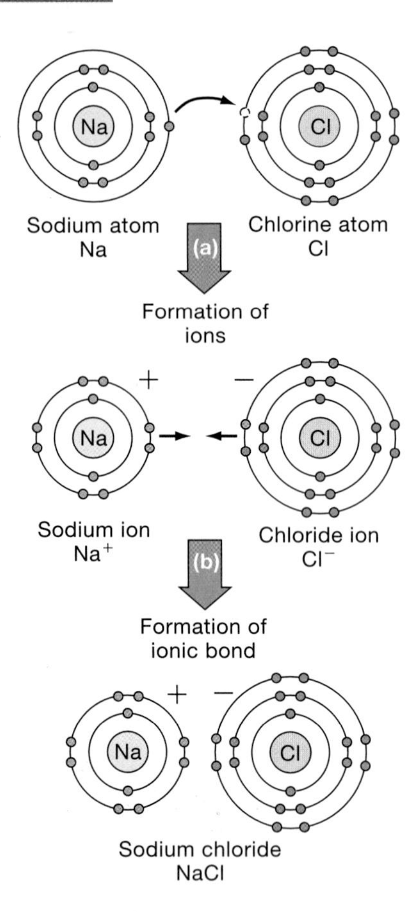

➢ Figure 2.2 **The formation of ions, or electrically charged atoms.** (a) When a neutral sodium atom loses the single electron in its outermost shell, the result is a sodium ion, Na^+. When a neutral chlorine atom gains an extra electron in its outer shell, the result is a chloride ion, Cl^-. (b) Oppositely charged ions attract one another. Such attraction creates an ionic bond and results in the formation of an ionic compound, in this case sodium chloride (NaCl).

it contains eight electrons. This principle, known as the **rule of octets,** is important for understanding chemical bonding, which we will discuss shortly.

Atoms whose outer electron shells are nearly full (containing six or seven electrons) or nearly empty (containing one or two electrons) tend to form ions. An **ion** is a charged atom produced when an atom gains or loses one or more electrons (➢ Figure 2.2a). When an atom of sodium (atomic number 11) loses the one electron in its outer shell without losing a proton, it becomes a positively charged ion, called a **cation** (kat'i-on). When an atom of chlorine (atomic number 17) gains an electron to fill its outer shell, it becomes a negatively charged ion, called an **anion** (an'i-on). In the ionized state, chlorine is referred to as chloride. Ions of elements such as sodium or chlorine are chemically more *stable* than atoms of these same elements because the ions' outer electron shells are full. Many elements are found in microorganisms or their environments as ions (Table 2.2). Those with one or two electrons in their outer shell tend to lose electrons and form ions with +1 or +2 charges, respectively; those with seven electrons in their outer shell tend to gain an electron and form ions with a charge of −1. Some ions, such as the hydroxyl (hi-drok'sil) ion (OH^-), are compounds—they contain more than one element.

Although all atoms of the same element have the same atomic number, they may not have the same atomic weight. **Atomic weight** is the sum of the number of protons and neutrons in an atom. Many elements consist of atoms with differing atomic weights. For example, carbon usually has six protons and six neutrons; it has an atomic weight of 12. But some naturally occurring carbon atoms have one or two extra neutrons, giving these atoms an atomic weight of 13 or 14. In addition, laboratory techniques are available to create atoms with different numbers of neutrons. Atoms of a particular element that contain different numbers of neutrons are called **isotopes.** The superscript to the left of the symbol for the element indicates the atomic weight of the particular isotope. For example, carbon with an atomic weight of 14, which is often used to date fossils, is written ^{14}C. The atomic weight of an element that has naturally occurring isotopes is the average atomic weight of the natural mixture of isotopes. Thus, atomic weights are not always whole numbers, even though any particular atom contains a specific number of whole neutrons and protons.

A **gram molecular weight,** or **mole,** is the weight of a substance in grams (g) equal to the sum of the atomic weights of the atoms in a molecule of the substance. For example, a mole of glucose, $C_6H_{12}O_6$, weighs 180 grams: [6 carbon atoms × 12 (atomic weight)] + [12 hydrogen atoms × 1 (atomic weight)] + [6 oxygen atoms × 16 (atomic weight)] = 180 grams. The mole is defined such that one mole of any substance always contains 6.023×10^{23} particles. Table 2.3 summarizes properties of elements found in living things.

➢ Table 2.2 **Some common ions**

Ion	Name	Brief description
Na^+	Sodium	Contributes to salinity of natural bodies of water and body fluids of multicellular organisms.
K^+	Potassium	Important ion that maintains cell turgor.
H^+	Hydrogen	Responsible for the acidity of solutions and commonly regulates motility.
Ca^{2+}	Calcium	Often acts as a chemical messenger.
Mg^{2+}	Magnesium	Commonly required for chemical reactions to occur.
Fe^{2+}	Ferrous iron	Carries electrons to oxygen during some chemical reactions that produce energy. Can prevent growth of some microbes that cause human disease.
NH_4^+	Ammonium	Found in animal wastes and degraded by some bacteria.
Cl^-	Chloride	Often found with a positively charged ion, where it usually neutralizes charge.
OH^-	Hydroxyl	Usually present in excess in basic solutions where H^+ is depleted.
HCO_3^-	Bicarbonate	Often neutralizes acidity of bodies of water and body fluids.
NO_3^-	Nitrate	A product of the action of certain bacteria that convert nitrite (NO_2^-) into a form plants can use.
SO_4^{2-}	Sulfate	Component of sulfuric acid in atmospheric pollutants and acid rain.
PO_4^{3-}	Phosphate	Can be combined with certain other molecules to form high-energy bonds where energy is stored in a form living things can use.

Nitrogen is called stickstoff in German and azoto in italian, but its symbol is N in every country of the world.

> Table 2.3 **Some properties of important elements found in living organisms (in order of abundance and importance)**

Element	Symbol	Atomic number	Atomic weight	Electrons in outer shell	Biological occurrence
Oxygen	O	8	16.0	6	Component of biological molecules; required for aerobic metabolism
Carbon	C	6	12.0	4	Essential atom of all organic compounds
Hydrogen	H	1	1.0	1	Component of biological molecules; H^+ released by acids
Nitrogen	N	7	14.0	5	Component of proteins and nucleic acids
Calcium	Ca	20	40.1	2	Found in bones and teeth; regulator of many cellular processes
Phosphorus	P	15	31.0	5	Found in nucleic acids, ATP, and some lipids
Sulfur	S	16	32.0	6	Found in proteins; metabolized by some bacteria
Iron	Fe	26	55.8	2	Carries oxygen; metabolized by some bacteria
Potassium	K	19	39.1	1	Important intracellular ion
Sodium	Na	11	23.0	1	Important extracellular ion
Chlorine	Cl	17	35.4	7	Important extracellular ion
Magnesium	Mg	12	24.3	2	Needed by many enzymes
Copper	Cu	29	63.6	1	Needed by some enzymes; inhibits growth of some microorganisms
Iodine	I	53	126.9	7	Component of thyroid hormones
Fluorine	F	9	19.0	7	Inhibits microbial growth
Manganese	Mn	25	54.9	2	Needed by some enzymes
Zinc	Zn	30	65.4	2	Needed by some enzymes; inhibits microbial growth

Some isotopes are stable, and others are not. The nuclei of unstable isotopes tend to emit subatomic particles and radiation. Such isotopes are said to be *radioactive* and are called **radioisotopes.** Emissions from radioactive nuclei can be detected by radiation counters. Such emissions can be useful in studying chemical processes, but they also can harm living things.

Chemical Bonds

Chemical bonds form between atoms through interactions of electrons in their outer shells. Energy associated with these bonding electrons holds the atoms together, forming molecules. Three kinds of chemical bonds commonly found in living organisms are ionic, covalent, and hydrogen bonds.

Ionic bonds result from the attraction between ions that have opposite charges. For example, sodium ions, with a positive charge (Na^+) combine with chloride ions, with a negative charge (Cl^-) (> Figure 2.2b).

Many compounds, especially those that contain carbon, are held together by **covalent bonds.** Instead of gaining or losing electrons as in ionic bonding, carbon and some other atoms in covalent bonds share pairs of electrons (> Figure 2.3). One carbon atom, which has four electrons in its outer shell, can share an electron with each of four hydrogen atoms. At the same time, each of the four hydrogen atoms shares an electron with the carbon atom. Four pair of electrons are shared, each pair consisting of one electron from carbon and one electron from hydrogen. Such mutual sharing makes a carbon atom stable with eight electrons in its outer shell and a hydrogen atom stable with two electrons in its outer shell. Equal sharing produces *nonpolar compounds*—compounds with no charged regions. Sometimes a carbon atom and an atom such as an oxygen atom share two pair of electrons to form a double bond. The octet rule still applies, and each atom has eight electrons in its outer shell and is therefore stable. In structural formulas, chemists use a single line to represent a single pair of shared electrons and a double line to represent two pair of shared electrons (> Figure 2.3).

Atoms of four elements, carbon, hydrogen, oxygen, and nitrogen, commonly form covalent bonds that fill their outer electron shells. Carbon shares four electrons, hydrogen one electron, oxygen two electrons, and nitrogen three electrons. Unlike many ionic bonds, covalent bonds are stable and thus are important in molecules that form biological structures.

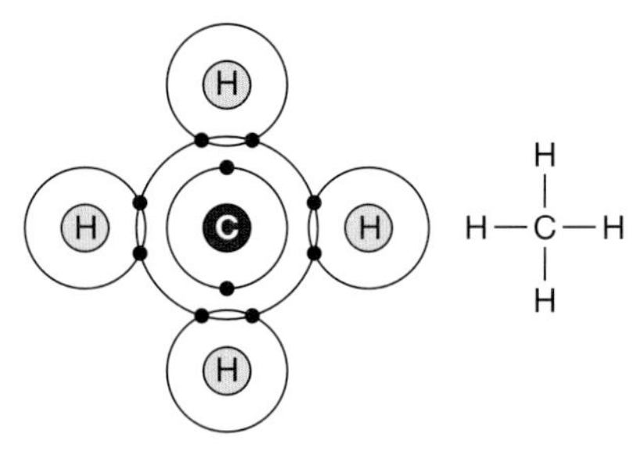

Covalent bonds: methane
(four single bonds)
CH_4

(a)

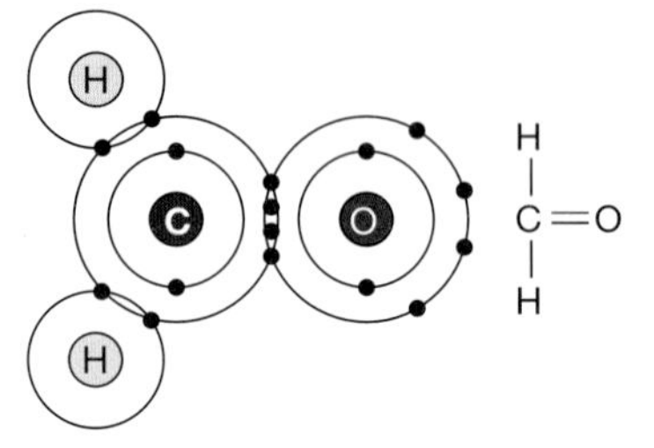

Formaldehyde
(double bond)
CH_2O

(b)

➢ Figure 2.3 **Covalent bonds are formed by sharing electrons.** (a) In methane, a carbon atom, with four electrons in its outermost shell, shares pairs of electrons with four hydrogen atoms. In this way all five atoms acquire stable, filled outer shells. Each shared electron pair constitutes a single covalent bond. (b) In formaldehyde, a carbon atom shares pairs of electrons with two hydrogen atoms and also shares two pairs of electrons with an oxygen atom, forming a double covalent bond.

Hydrogen bonds, though weaker than ionic and covalent bonds, are important in biological structures and are typically present in large numbers. The atomic nuclei of oxygen and nitrogen attract electrons very strongly. When hydrogen is covalently bonded to oxygen or nitrogen, the electrons of the covalent bond are shared unevenly—they are held closer to the oxygen or nitrogen than to the hydrogen. The hydrogen atom then has a partial positive charge, and the other atom has a partial negative charge. In this case of unequal sharing, the molecule is called a **polar compound** because of its oppositely charged regions. The weak attraction between such partial charges is called a hydrogen bond.

Polar compounds such as water often contain hydrogen bonds. In a water molecule, electrons from the hydrogen atoms stay closer to the oxygen atom, and the hydrogen atoms lie to one side of the oxygen atom (➢ Figure 2.4). Thus, water molecules are polar molecules that have a positive hydrogen region and a negative oxygen region. Covalent bonds between the hydrogen and oxygen atoms hold the atoms together. Hydrogen bonds between the hydrogen and oxygen regions of different water molecules hold the molecules in clusters.

Hydrogen bonds also contribute to the structure of large molecules such as proteins and nucleic acids, which contain long chains of atoms. The chains are coiled or folded into a three-dimensional configuration that is held together in part by hydrogen bonds.

Chemical Reactions

Chemical reactions in living organisms typically involve the use of energy to form chemical bonds and the release of energy as chemical bonds are broken. For example, the food we eat consists of molecules that have much energy stored in their chemical bonds. During **catabolism** (ka-tab'o-lizm), the breakdown of substances, food is degraded and some of that stored energy is released. Microorganisms use nutrients in the same general way. A catabolic reaction can be symbolized as follows:

$$X—Y \rightarrow X + Y + \text{energy}$$

where X—Y represents a nutrient molecule and where energy was originally stored in the bond between X and Y.

Catabolic reactions are **exergonic**—that is, they release energy. Conversely, energy is used to form chemical bonds in the synthesis of new compounds. In **anabolism** (a-nab'o-lizm), the buildup, or *synthesis,* of substances, energy is used to create bonds. An anabolic reaction can be symbolized as follows:

$$X + Y + \text{energy} \rightarrow X—Y$$

where energy is stored in the new substance X—Y. Anabolic reactions occur in living cells when small molecules are used to synthesize large molecules. Cells can store small amounts of

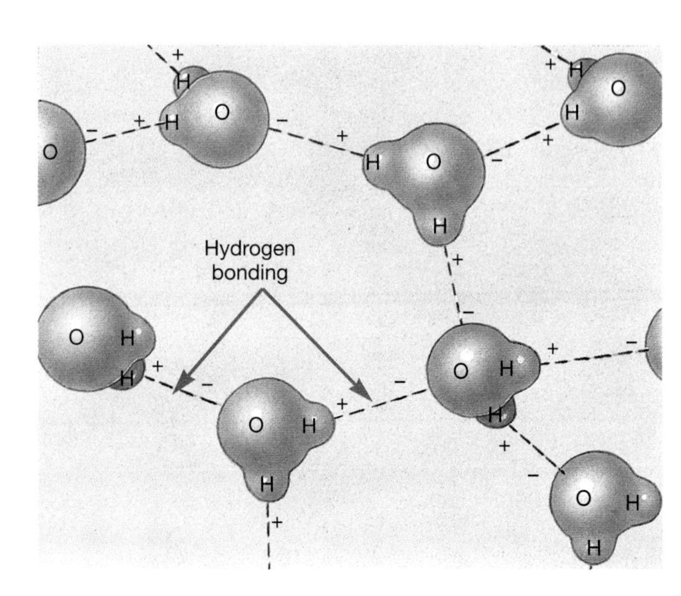

➢ Figure 2.4 **Polar compounds and hydrogen bonding.** Water molecules are polar—they have a region with a partial positive charge (the hydrogen atoms) and a region with a partial negative charge (the oxygen atom). Hydrogen bonds, created by the attraction between oppositely charged regions of different molecules, hold the water molecules together in clusters.

energy for later use or can expend energy to make new molecules. Most anabolic reactions are **endergonic**—that is, they require energy.

- ✓ Which number tells us the identity of an atom?
- ✓ Is it possible to have a molecule of an element? A molecule of a compound? Give examples.
- ✓ If isotopes can be thought of as "twins," "triplets," "sextuplets," and so on, in what ways are they identical? Different?
- ✓ What type of bonding is produced by the equal sharing of an electron pair between two atoms? By unequal sharing?

Water and Solutions

Water, one of the simplest of chemical compounds, is also one of the most important to living things. It takes part directly in many chemical reactions. Numerous substances dissolve in water or form mixtures called colloidal dispersions. Acids and bases exist and function principally in water mixtures.

Water

Water is so essential to life that humans can live only a few days without it. Many microorganisms die almost immediately if removed from their normal aqueous environments, such as lakes, ponds, oceans, and moist soil. Yet, others can survive for several hours or days without water, and spores formed by a few microorganisms survive for many years away from water. Several bacteria find the moist, nutrient-rich secretions of human skin glands to be an ideal environment.

Water has several properties that make it important to living things. Because water is a polar compound and forms hydrogen bonds, it can form thin layers on surfaces and can act as a *solvent,* or dissolving medium. Water is a good solvent for ions because the polar water molecules surround the ions. The positive region of water molecules is attracted to negative ions, and the negative region of water molecules is attracted to positive ions. Many different kinds of ions can therefore be distributed evenly through a water medium, forming a *solution* (➢ Figure 2.5).

Water forms thin layers because it has a high surface tension. **Surface tension** is a phenomenon in which the surface of water acts as a thin, invisible, elastic membrane (➢ Figure 2.6). The polarity of water molecules gives them a strong attraction for one another but no attraction for gas molecules in air at the water's surface. Therefore, surface water molecules cling together, forming hydrogen bonds with other molecules below the surface. In living cells this feature of surface tension allows a thin film of water to cover membranes and to keep them moist.

Water has a high *specific heat,* that is, it can absorb or release large quantities of heat energy with little temperature change. This property of water helps to stabilize the temperature of living organisms, which are composed mostly of water, as well as bodies of water where many microorganisms live.

Finally, water provides the medium for most chemical reactions in cells, and it participates in many of these reactions. Suppose, for example, that substance X can gain or lose H^+ and that

A Winter Dilemma

Picture it—a harsh winter, snow and ice, roads covered with salt and sand. Now it's spring. However, there's something wrong with the trees and other plants growing along the roads. Has excessive release of chemicals from winter salt runoff affected soil chemistry or the soil's ability to support microbes?

As an environmental scientist with the State Laboratories, it's your job to investigate. You test the soil and runoff water for chemicals used during the winter. Are there high concentrations of de-icing chemicals near the affected roads? How far have they spread? Where is normal plant growth again observed? Do collected soils demonstate typical and divergent microbial populations? Are there any unusual pathologies associated with the plants growing near the affected area?

Place yourself in this scenario and use the scientific method to design an experiment to show what is happening here.

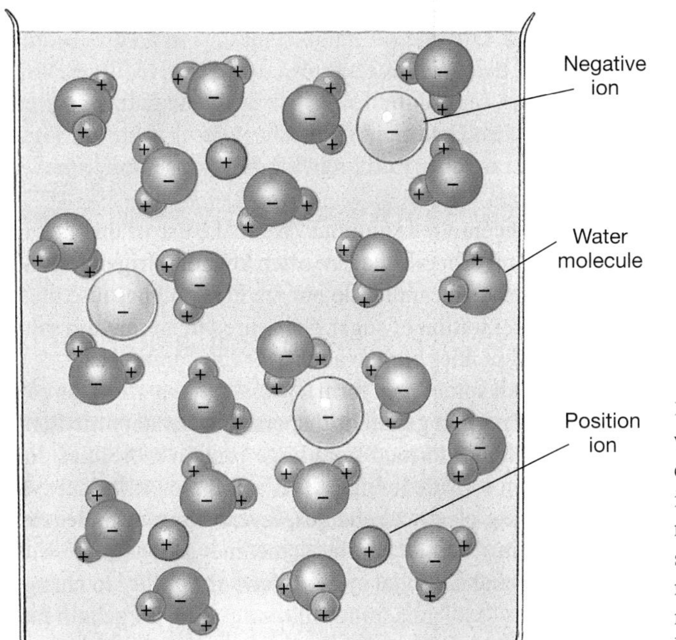

➢ Figure 2.5 **Polarity and water molecules.** Polarity enables water to dissolve many ionic compounds. The positive regions of the water molecules surround negative ions, and the negative regions of the water molecules surround positive ions, holding the ions in solution.

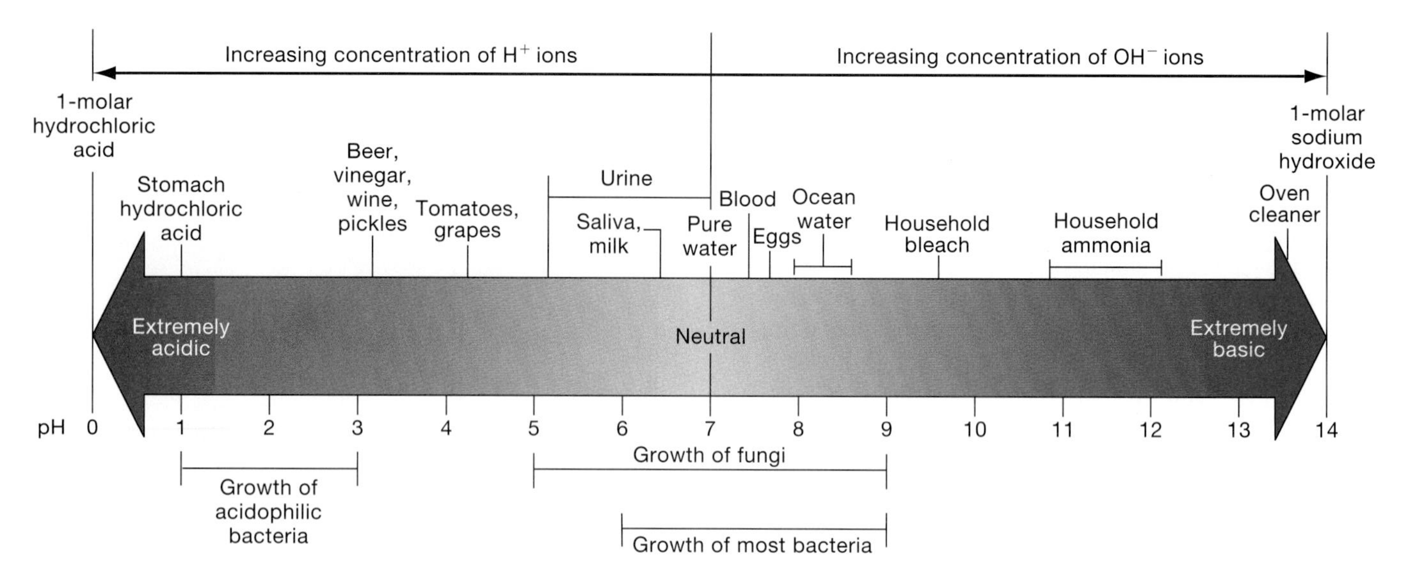

➢ Figure 2.7 **The pH values of some common substances.** Each unit of the pH scale represents a tenfold increase or decrease in the concentration of hydrogen ions. Thus, vinegar, for example, is 10,000 times more acidic than pure water.

✓ What properties of a water molecule enable it to act as a good solvent for ionic molecules?

✓ Why is a higher number, such as pH 11, a stronger base than pH 9, whereas a higher number, such as pH 5, is a weaker acid than pH 3?

Chemists have devised the concept of pH to specify the acidity or alkalinity of a solution. The **pH** scale (➢ Figure 2.7), which relates proton concentrations to pH, is a logarithmic scale. This means that the concentration of hydrogen ions (protons) changes by a factor of 10 for each unit of the scale. The practical range of the pH scale is from 0 to 14. A solution with a pH of 7 is **neutral**—neither acidic nor **alkaline** (basic). Pure water has a pH of 7 because the concentrations of H^+ and OH^- in it are equal. ➢ Figure 2.7 shows the pH of some body fluids, selected foods, and other substances.

Some Like It Sour

Most natural environments have pH values between 5 and 9, and microbes living in those locations grow within that pH range. The known species of bacteria that grow at pH values below 2 or greater than 10 often have unique properties that we are able to use to our advantage. Certain bacteria that live at low pH, acidophiles, are used to leach economically important metals from low-yielding ores. Low-grade copper ore, dumped into a large pile, is treated with a liquid containing dilute sulfuric acid and the bacterium *Thiobacillus ferrooxidans*. The presence of *Thiobacillus ferrooxidans* increases the rate of copper sulfite oxidation and the production of copper sulfate. Because copper sulfate is extremely water-soluble, it is possible to economically extract and precipitate copper from the liquid collected from the leach dump.

Complex Organic Molecules

The basic principles of general chemistry also apply to **organic chemistry,** the study of compounds that contain carbon. The study of the chemical reactions that occur in living systems is the branch of organic chemistry known as **biochemistry.** Early in the 1800s it was believed that molecules from living things were filled with a supernatural "vital force" and therefore could not be explained by the laws of chemistry and physics. It was considered impossible to make *organic compounds* outside of living systems. That idea was disproved in 1828 when the German scientist Friedrich Wohler synthesized the organic compound urea, a small molecule excreted as a waste material by many animals. Since that time thousands of organic compounds—plastics, fertilizers, and medicines—have been made in the laboratory. Organic compounds such as carbohydrates, lipids, proteins, and nucleic acids occur naturally in living things and in the products or remains of living things. The ability of carbon atoms to form covalent bonds and to link up in long chains makes possible the formation of an almost infinite number of organic compounds.

The simplest carbon compounds are the *hydrocarbons,* chains of carbon atoms with their associated hydrogen atoms. The structure of the hydrocarbon propane, C_3H_8, for example, is as follows:

```
    H   H   H
    |   |   |
H — C — C — C — H
    |   |   |
    H   H   H
```

Carbon chains can have not only hydrogen but other atoms such as oxygen and nitrogen bound to them. Some of these atoms form functional groups. A **functional group** is a part of a molecule that generally participates in chemical reactions as a unit and that gives the molecule some of its chemical properties.

Ketone

Alcohol

Aldehyde

Organic acid

Reduced ⟷ Oxidized

➢ Figure 2.8 **Four classes of organic compounds that incorporate oxygen.** Alcohols contain one or more hydroxyl groups (—OH); aldehydes and ketones contain carbonyl groups (—C=O); and organic acids contain carboxyl groups (—COOH).

Four significant groups of compounds-alcohols, aldehydes, ketones, and organic acids—have functional groups that contain oxygen (➢ Figure 2.8). An alcohol has one or more hydroxyl groups (— OH). An aldehyde has a carbonyl group (— CO) at the end of the carbon chain; a ketone has a carbonyl group within the chain. An organic acid has one or more carboxyl groups (—COOH). One key functional group that does not contain oxygen is the amino group (— NH_2). Found in amino acids, amino groups account for the nitrogen in proteins.

The relative amount of oxygen in different functional groups is significant. Groups with little oxygen, such as alcohol groups, are said to be *reduced;* groups with relatively more oxygen, such as carboxyl groups, are said to be *oxidized* (➢ Figure 2.8). As we shall see in Chapter 5, *oxidation* is the addition of oxygen or the removal of hydrogen or electrons from a substance. Burning is an example of oxidation. *Reduction* is the removal of oxygen or the addition of hydrogen or electrons to a substance. In general, the more reduced a molecule, the more energy it contains. Hydrocarbons, such as gasoline, have no oxygen and thus represent the extreme in energy-rich, reduced molecules. They make good fuels because they contain so much energy. Conversely, the more oxidized a molecule, the less energy it contains. Carbon dioxide (CO_2) represents the extreme in an oxidized molecule because no more than two oxygen atoms can bond to a single carbon atom. As we shall see, oxidation releases energy from molecules.

Let us now consider the major classes of large, complex biochemical molecules of which all living things, including microbes, are composed.

Carbohydrates

Carbohydrates serve as the main source of energy for most living things. Plants make carbohydrates, including structural carbohydrates such as cellulose and energy-storage carbohydrates such as starch. Animals use carbohydrates as food, and many, including humans, store energy in a carbohydrate called *glycogen*. Many microorganisms use carbohydrates from their environment for energy and also make a variety of carbohydrates. Carbohydrates in the membranes of cells can act as markers that make a cell chemically recognizable. Chemical recognition is important in immunological reactions and other processes in living things.

All carbohydrates contain the elements carbon, hydrogen, and oxygen, generally in the proportion of two hydrogen atoms to each carbon and oxygen atom. There are three groups of carbohydrates: monosaccharides, disaccharides, and polysaccharides. **Monosaccharides** consist of a carbon chain or ring with several alcohol groups and one other functional group, either an aldehyde group or a ketone group. Several monosaccharides, such as glucose and fructose, are **isomers**—they have the same molecular formula, $C_6H_{12}O_6$, but different structures and different properties (➢ Figure 2.9). Thus, even at the chemical level we can see that structure and function are related.

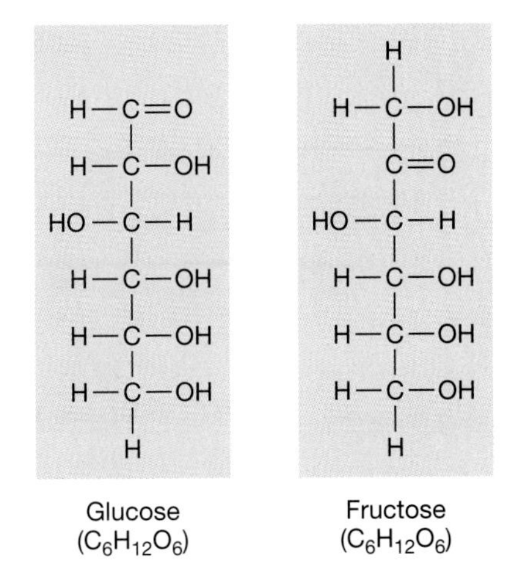

➢ Figure 2.9 **Isomers.** Glucose and fructose are isomers: They contain the same atoms, but they differ in structure.

Glucose, the most abundant monosaccharide, can be represented schematically as a straight chain, a ring, or a three-dimensional structure. The chain structure, in ➢ Figure 2.10a, clearly shows a carbonyl group at carbon 1 (the first carbon in the chain, at the top in this orientation)

➢ Figure 2.10 **Three ways of representing the glucose molecule.** (a) In solution, the straight-chain form is rarely found. (b) Instead, the molecule bonds to itself, forming a six-membered ring. The ring is conventionally depicted as a flat hexagon. (c) The actual three-dimensional structure is more complex. The spheres in this depiction represent carbon atoms.

> **Candy-Coated Chemistry**
>
> Liquid centers, such as those found in chocolate-covered cherrries, are made from invert sugar. Invert sugar is used to make many different confections including jelly, candy, and prepared frostings. Invert sugar is beet sugar (sucrose) that has been treated with the enzyme *invertase.* The enzyme breaks down sucrose into its two components, glucose and fructose, thereby reducing the size of the sugar crystals. Invert sugar produces a smoother product as a result of its fine crystal structure. In nature, invertase is associated with many tuber vegetables, such as beets and potatoes, and is synthesized by many bacteria. What do you think the role of invertase might be in nature? in bacteria?

and alcohol groups on all the other carbons. ➢ Figure 2.10b shows how a glucose molecule in solution rearranges and bonds to itself to form a closed ring. The three-dimensional projection in ➢ Figure 2.10c more closely approximates the actual shape of the molecule. In studying structural formulas, it is important to imagine each molecule as a three-dimensional object.

Monosaccharides can be reduced to form deoxy sugars and sugar alcohols (➢ Figure 2.11). The deoxy sugar *deoxyribose,* which has a hydrogen atom instead of an —OH group on one of its carbons, is a component of DNA. Certain sugar alcohols, which have an additional alcohol group instead of an aldehyde or ketone group, can be metabolized by particular microorganisms. Mannitol and other sugar alcohols are used to identify some microorganisms in diagnostic tests.

Disaccharides are formed when two monosaccharides are connected by the removal of water and the formation of a **glycosidic bond,** a sugar alcohol/sugar linkage (➢ Figure 2.12a). Sucrose, common table sugar, is a disaccharide made of glucose and fructose. **Polysaccharides** are formed when many monosaccharides are linked by glycosidic bonds (➢ Figure 2.12b). Polysaccharides such as starch, glycogen, and cellulose are **polymers**—long chains of repeating units—of glucose. However, the glycosidic bonds in each polymer are arranged differently. Plants and most algae make starch and cellulose. Starch serves as a way to store energy, and cellulose is a structural component of cell walls. Animals make and store glycogen, which they can break down to glucose as energy is needed. Microorganisms contain several other important polysaccharides, as we shall see in later chapters.

Table 2.4 summarizes the types of carbohydrates.

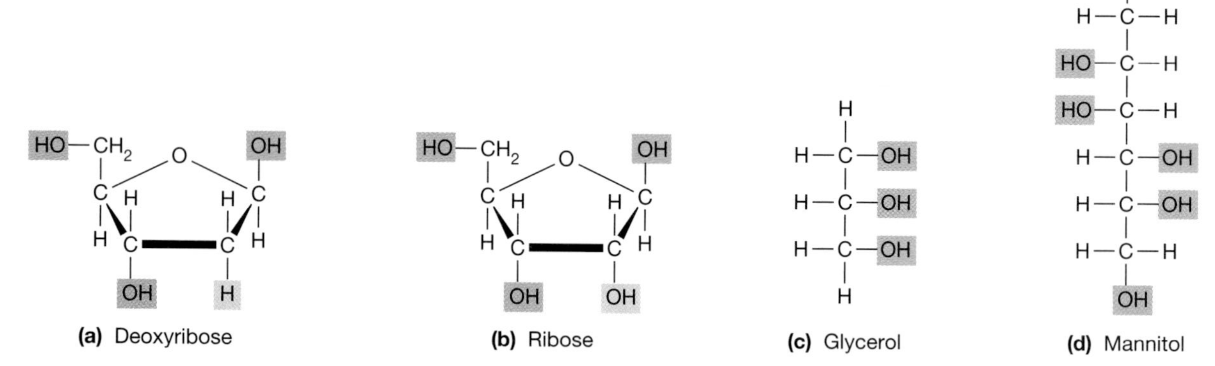

➢ Figure 2.11 **Deoxy sugars and sugar alcohols.** (a) "Deoxy" indicates one less oxygen atom. One of the carbon atoms of the deoxy sugar deoxyribose lacks a hydroxyl group that (b) ribose has. (c) Glycerol is a three-carbon sugar alcohol that is a component of fats. (d) Mannitol is a sugar alcohol used in diagnostic tests for certain microbes.

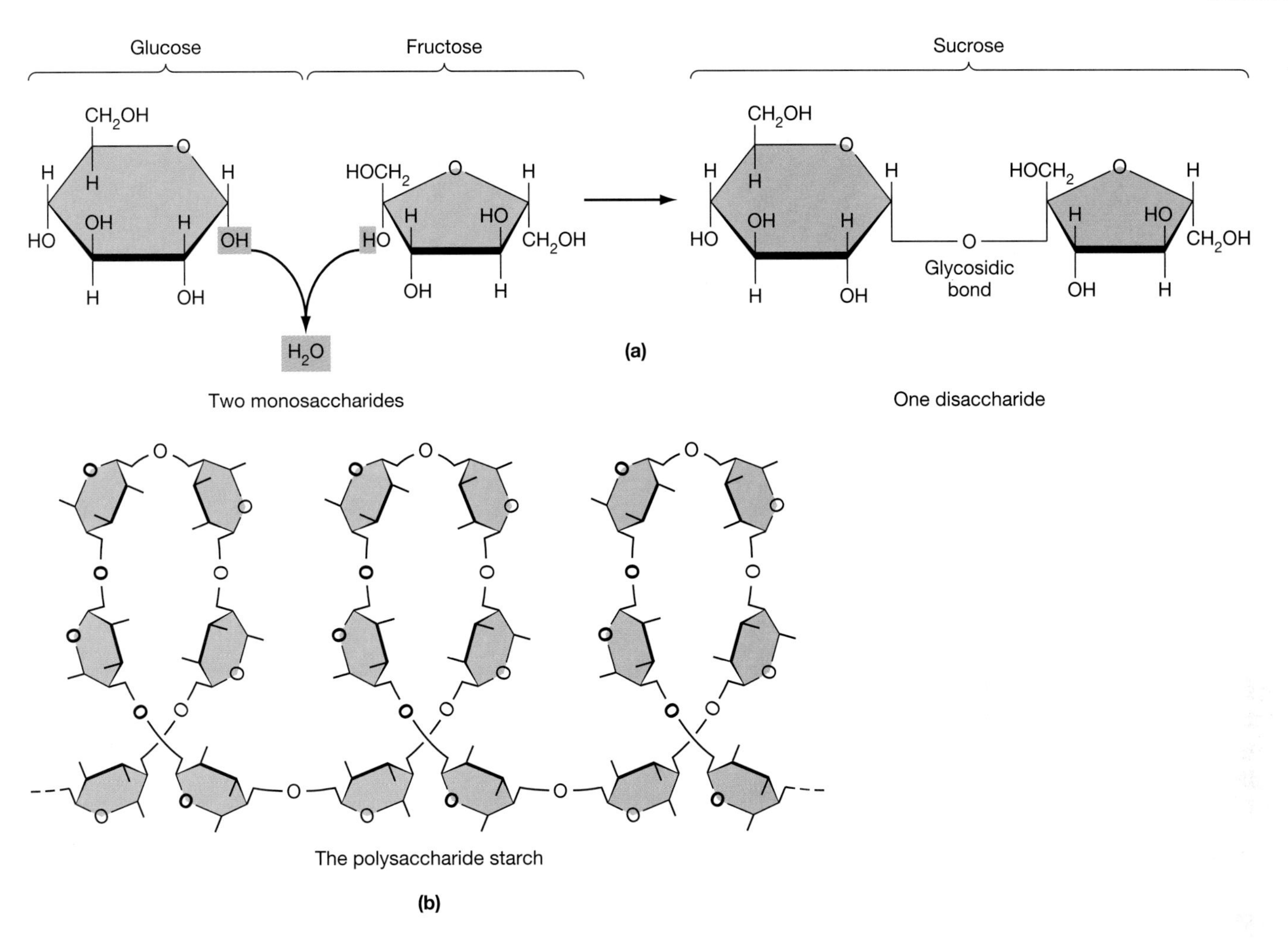

➢ Figure 2.12 **Disaccharides and polysaccharides.** (a) Two monosaccharides are joined to form a disaccharide by dehydration synthesis and the formation of a glycosidic bond. (b) Polysaccharides such as starch are formed by similar reactions that link many monosaccharides into long chains.

➢ Table 2.4 **Types of carbohydrates**

Class of carbohydrates	Examples	Description and occurrence
Monosaccharides	Glucose	Sugar found in most organisms
	Fructose	Sugar found in fruit
	Galactose	Sugar found in milk
	Ribose	Sugar found in RNA
	Deoxyribose	Sugar found in DNA
Disaccharides	Sucrose	Glucose and fructose; table sugar
	Lactose	Glucose and galactose; milk sugar
	Maltose	Two glucose units; product of starch digestion
Polysaccharides	Starch	Polymer of glucose stored in plants, digestible by humans
	Glycogen	Polymer of glucose stored in animal liver and skeletal muscles
	Cellulose	Polymer of glucose found in plants, not digestible by humans; digested by some microbes

Ticks' body walls are impervious to most pesticides thanks to chitin, a polysaccharide, also found in fungal cell walls.

Can a Cow Actually Explode?

Cows can derive a good deal of nourishment from grass, hay, and other fibrous vegetable matter that are inedible to humans. We can't digest cellulose, the chief component of plants. If you had to live on hay, you would probably starve to death. How, then, do cows and other hooved animals manage on such a diet?

Oddly enough, cows can't digest cellulose either. But they don't need to—it's done for them. Cows and their relatives harbor in their stomachs large populations of microorganisms that do the work of breaking down cellulose into sugars that the animal can use. The same is true of termites: If it weren't for microbes in their guts that help them to digest cellulose, they couldn't dine on the wooden beams in your house.

Cellulose is very similar to starch—both consist of long chains of glucose molecules. The bonds between these molecules, however, are slightly different in their geometry in the two substances. As a result, the enzymes that animals use to break down a starch molecule into its component glucose units have no effect on cellulose. In fact, very few organisms produce enzymes that can attack cellulose. Even the protists (unicellular organisms with a nucleus) that live in the stomachs of cows and termites cannot always do it by themselves. Just as cows and termites depend on the protists in their stomachs, the protists frequently depend on certain bacteria that reside permanently within them. It is these bacteria that actually make the essential digestive enzymes.

The activities of the intestinal microorganisms that perform these digestive services are a mixed blessing, both to the cows and to the humans who keep them. The bacteria also produce methane gas, CH_4—as much as 190 to 380 l per day from a single cow. (Methane production can be so rapid that a cow's stomach may rupture if the cow can't burp. Some ingenious inventors have actually patented cow safety valves to release the gas buildup directly through the animal's side.) When this gas eventually makes its way out of the cow by one route or another, it rises to the upper atmosphere. There it is suspected of contributing to the "greenhouse effect," trapping solar heat and causing an overall warming of the earth's climate (Chapter 25). Scientists have estimated that the world's cows release 50 million metric tons of methane annually; that's not counting the sheep, goats, antelope, water buffalo, and other grass eaters.

Higher fat content of food causes slower movement of feces through the bowels. There, bacteria convert the undigested fats into cancer-causing compounds.

Evidence from Babylonian excavations indicate that soapmaking was known as early as 2800 B.C.

Lipids

Lipids constitute a chemically diverse group of substances that includes fats, phospholipids, and steroids. They are relatively insoluble in water but are soluble in nonpolar solvents such as ether and benzene. Lipids form part of the structure of cells, especially cell membranes, and many can be used for energy. Generally, lipids contain relatively more hydrogen and less oxygen than carbohydrates and therefore contain more energy than carbohydrates.

Fats contain the three-carbon alcohol glycerol and one or more fatty acids. A **fatty acid** consists of a long chain of carbon atoms with associated hydrogen atoms and a carboxyl group at one end of the chain. The synthesis of a fat from glycerol and fatty acids involves removing water and forming an ester bond between the carboxyl group of the fatty acid and an alcohol group of glycerol (➢ Figure 2.13a). A **triacylglycerol,** formerly called a *triglyceride,* is a fat formed when three fatty acids are bonded to glycerol. *Monoacylglycerols* (monoglycerides) and *diacylglycerols* (diglycerides) contain one and two fatty acids, respectively, and usually are formed from the digestion of triacylglycerols.

Fatty acids can be saturated or unsaturated. A **saturated fatty acid** contains all the hydrogen it can have; that is, it is saturated with hydrogen (➢ Figure 2.13b). An **unsaturated fatty acid** has lost at least two hydrogen atoms and contains a double bond between the two carbons that have lost hydrogen atoms (➢ Figure 2.13c). "Unsaturated" thus means not completely saturated with hydrogen. Oleic acid is an unsaturated fatty acid. *Polyunsaturated fats,* many of which are vegetable oils that remain liquid at room temperature, contain many unsaturated fatty acids.

Some lipids contain one or more other molecules in addition to fatty acids and glycerol. For example, **phospholipids,** which are found in all cell membranes, differ from fats by the substitution of phosphoric acid (H_3PO_4) for one of the fatty acids (➢ Figure 2.14a). The charged phosphate group (— HPO_4^-) typically attaches to another charged group. Both can mix with

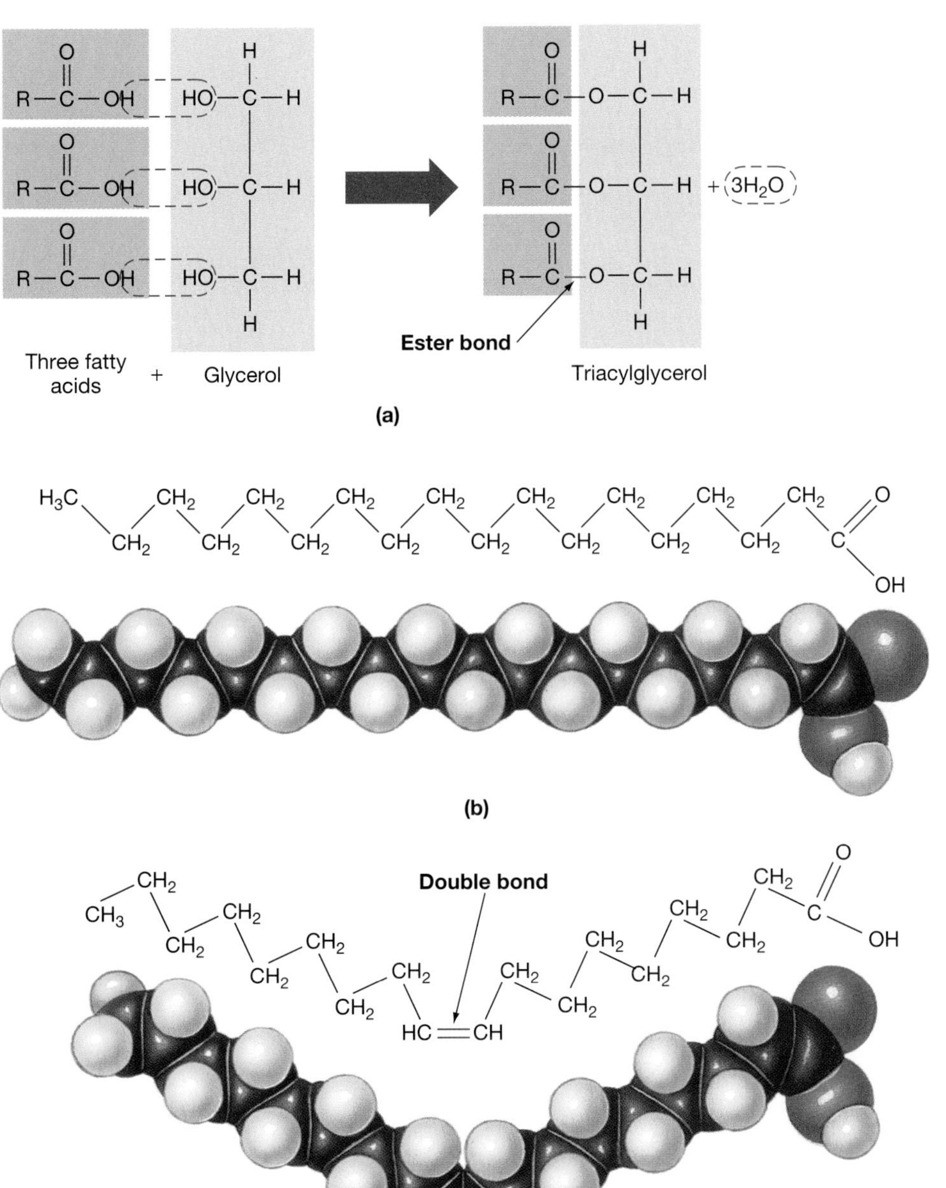

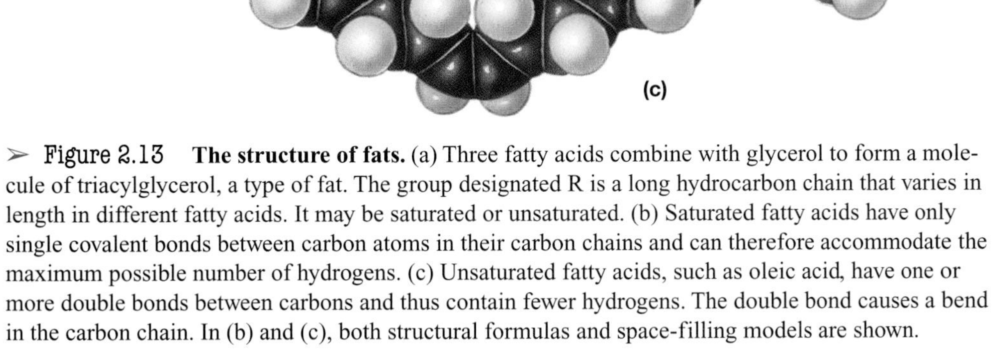

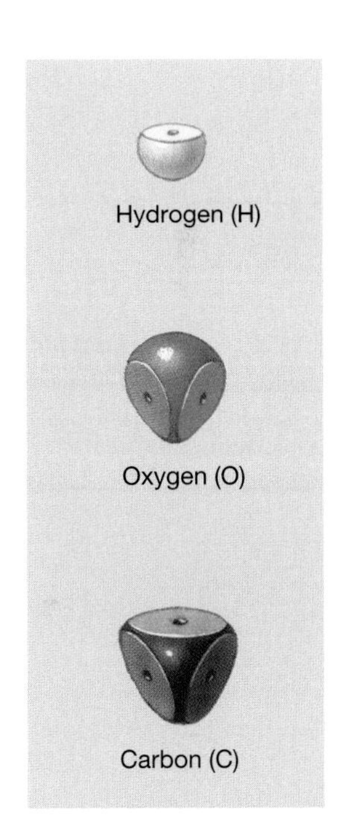

➢ Figure 2.13 **The structure of fats.** (a) Three fatty acids combine with glycerol to form a molecule of triacylglycerol, a type of fat. The group designated R is a long hydrocarbon chain that varies in length in different fatty acids. It may be saturated or unsaturated. (b) Saturated fatty acids have only single covalent bonds between carbon atoms in their carbon chains and can therefore accommodate the maximum possible number of hydrogens. (c) Unsaturated fatty acids, such as oleic acid, have one or more double bonds between carbons and thus contain fewer hydrogens. The double bond causes a bend in the carbon chain. In (b) and (c), both structural formulas and space-filling models are shown.

water, but the fatty acid end cannot (➢ Figure 2.14b). Such properties of phospholipids are important in determining the characteristics of cell membranes (Chapter 4).

Steroids have a four-ring structure (➢ Figure 2.15a) and are quite different from other lipids. They include cholesterol, steroid hormones, and vitamin D. Cholesterol (➢ Figure 2.15b) is insoluble in water and is found in the cell membranes of animal cells and the group of bacteria called mycoplasmas. Steroid hormones and vitamin D are important in many animals.

Soap's effectiveness is reduced in hard water when mineral salts, like calcium, magnesium, and iron, react with the soap and form an insoluble precipitate (soap scum).

Uncharged fatty acid chains

Other charged group

Charged phosphate group

Glycerol portion

(a)

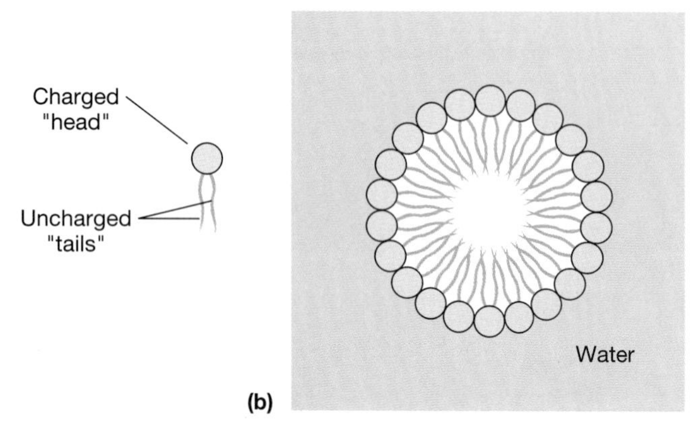

➢ Figure 2.14 **Phospholipids.** (a) In phospholipids, one of the fatty acid chains of a fat molecule is replaced by phosphoric acid. The charged phosphate group and another attached group can interact with water molecules, which are polar, but the two long, uncharged fatty acid tails cannot. (b) As a result, phospholipid molecules in water tend to form globular structures with the phosphate groups facing outward and the fatty acids in the interior.

Down the Drain

You have washed enough dishes to know that soaps get oily substances off the dishes and into the water. Although you probably don't think about how soap works, soap chemistry is an application of what you have learned in Chapter 2. Because water is a polar molecule, it has a high surface tension and beads on clean surfaces. To make water "wetter" it is necessary to reduce the surface tension by adding surfactants. Soaps are anionic surfactants made from fats and oils treated with strong alkali. This process produces a complex molecule with a charged carboxyl group at one end and a nonionized saturated hydrocarbon at the other. The saturated hydrocarbon portion of the soap molecule mixes with the fats on dishes while the charged carboxyl group mixes with the dishwater. The chemical interaction between the grease, soap, and water loosens the food particles off your dishes and down the drain.

Side chain

(a) (b)

➢ Figure 2.15 **Steroids.** (a) Steroids are lipids with a characteristic four-ring structure. The specific chemical groups attached to the rings determine the properties of different steroids. (b) One of the most biologically important steroids is cholesterol, a component of the membranes of animal cells and one group of bacteria.

Proteins

Properties of Proteins and Amino Acids

Among the molecules found in living things, proteins have the greatest diversity of structure and function. **Proteins** are composed of building blocks called **amino acids,** which have at least one amino (—NH_2) group and one acidic carboxyl (—COOH) group. The general structure of an amino acid and some of the 20 amino acids found in proteins are shown in ➢ Figure 2.16. Each amino acid is distinguishable by a different chemical group, called an **R group,** attached to the central carbon atom. Because all amino acids contain carbon, hydrogen, oxygen, and nitrogen, and some contain sulfur, proteins also contain these elements.

A protein is a polymer of amino acids joined by **peptide bonds**—that is, covalent bonds that link an amino group of one amino acid and a carboxyl group of another amino acid (➢ Figure 2.17). Two amino acids linked together make a *dipeptide,* three make a *tripeptide,* and many make a **polypeptide.** In addition to the amino and carboxyl groups, some amino acids have an R group called a *sulfhydryl group* (—SH). Sulfhydryl groups in adjacent chains of amino acids can lose hydrogen and form *disulfide linkages* (—S—S—) from one chain to the other.

The Structure of Proteins

Proteins have several levels of structure. The **primary structure** of a protein consists of the specific sequence of amino acids in a polypeptide chain (➢ Figure 2.18a). The **secondary structure** of a protein consists of the folding or coiling of amino acid chains into a particular pattern, such as a helix or pleated sheet (➢ Figure 2.18b). Hydrogen bonds are responsible for such patterns. Further bending and folding of the protein molecule into globular (irregular spherical) shapes or fibrous threadlike strands produces the **tertiary structure** (➢ Figure 2.18c). Some large proteins such as hemoglobin have **quaternary structure,** formed by the association of several tertiary-structured polypeptide chains (➢ Figure 2.19). Tertiary and quaternary structures are maintained by disulfide linkages, hydrogen bonds, and other forces between R groups of amino acids. The three-dimensional shapes of protein molecules and the nature of sites at which other molecules can bind to them are extremely important in determining how proteins function in living organisms.

Several conditions can disrupt hydrogen bonds and other weak forces that maintain protein structure. They include highly acidic or basic conditions and temperatures above 50°C. Such disruption of secondary, tertiary, and quatenary structures is called **denaturation.** Sterilization and disinfection procedures often make use of heat or chemicals that kill microorganisms by denaturing their proteins. Also, the cooking of meat tenderizes it by denaturing proteins. Therefore, microbes and cells of larger organisms must be maintained within fairly narrow ranges of pH and temperature to prevent disruption of protein structure.

Classification of Proteins

Most proteins can be classified by their major functions as structural proteins or enzymes. **Structural proteins,** as the name implies, contribute to the three-dimensional structure of cells, cell parts, and membranes. Certain proteins, called *motile proteins,* contribute both to

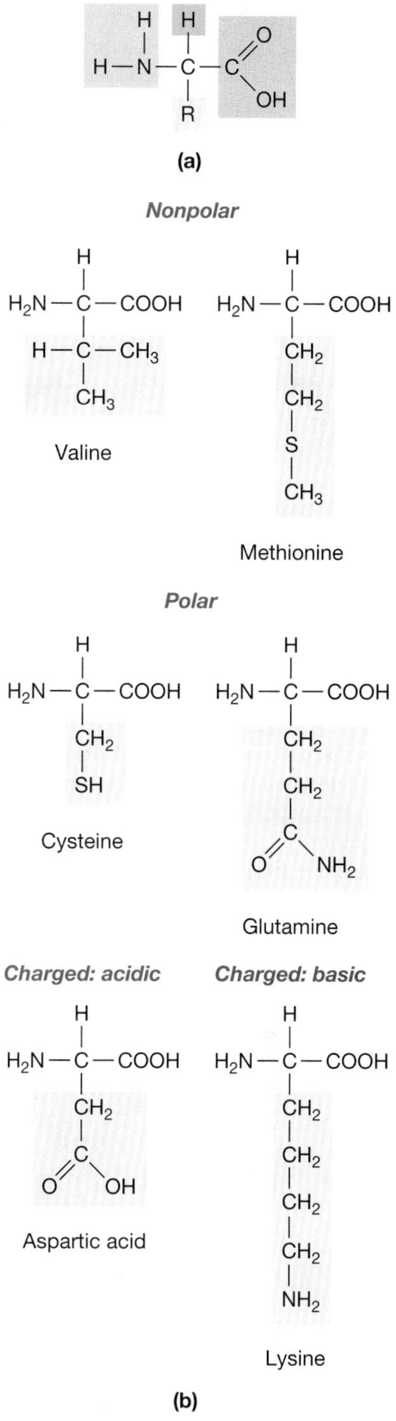

(a)

Nonpolar

Valine

Methionine

Polar

Cysteine

Glutamine

Charged: acidic

Aspartic acid

Charged: basic

Lysine

(b)

➢ Figure 2.16 **Amino acids.** (a) The general structure of an amino acid, and (b) six representative examples. All amino acids have four groups attached to the central carbon atom: an amino (—NH_2) group, a carboxyl (—COOH) group, a hydrogen atom, and a group designated R that is different in each amino acid. The R group determines many of the chemical properties of the molecule—for example, whether it is nonpolar, polar, acidic, or basic.

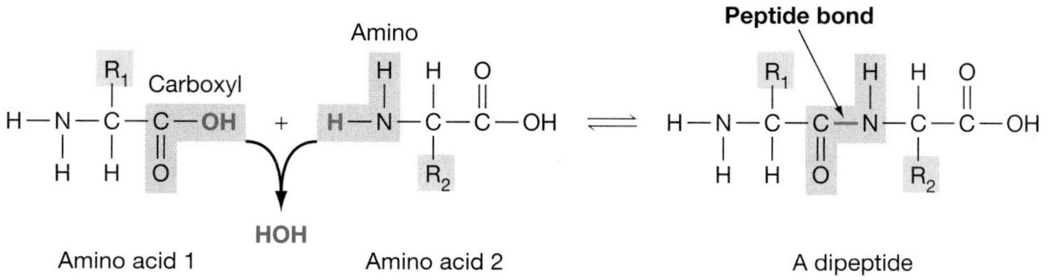

Amino

Carboxyl

Peptide bond

HOH

Amino acid 1

Amino acid 2

A dipeptide

➢ Figure 2.17 **Peptide linkage.** Two amino acids are joined by the removal of a water molecule (dehydration synthesis) and the formation of a peptide bond between the —COOH group of one and the —NH_2 group of the other.

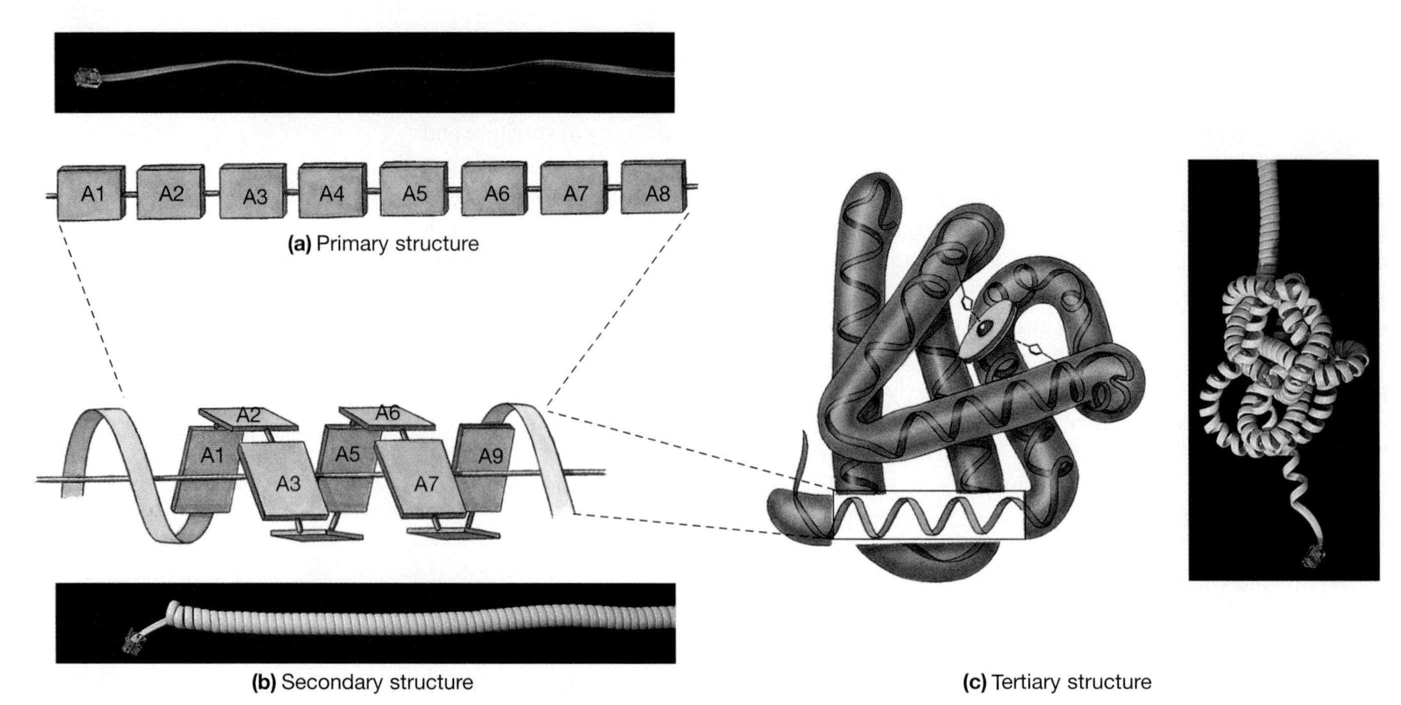

➢ Figure 2.18 **Three levels of protein structure.** (a) Primary structure is the sequence of amino acids (A1, A2, . . .) in a polypeptide chain. Imagine it as a straight telephone cord. (b) Polypeptide chains, especially those of structural proteins, tend to coil or fold into a few simple, regular, three-dimensional patterns called secondary structure. Imagine the telephone cord as a coiled cord. (c) Polypeptide chains of enzymes and other soluble proteins may also exhibit secondary structure. In addition, the chains tend to fold up into complex, globular shapes that constitute the protein's tertiary structure. Imagine the knot formed when a coiled telephone cord tangles.

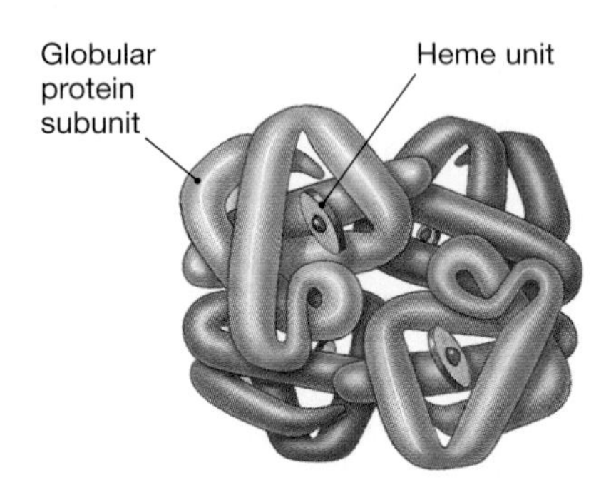

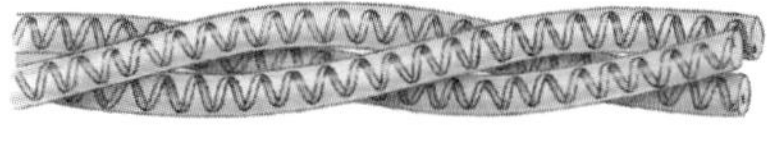

➢ Figure 2.19 **Quaternary protein structure.** (a) Many large proteins such as hemoglobin, which carries oxygen in human red blood cells, are made up of several polypeptide chains. The arrangement of these chains makes up the protein's quaternary structure. (b) Some structural proteins such as keratin, a component of human skin and hair, also consist of several polypeptide chains and so have quaternary structure.

structure and to movement. They account for the contraction of animal muscle cells and for some kinds of movement in microbes. **Enzymes** are protein *catalysts*—substances that control the rate of chemical reactions in cells. A few proteins are neither structural proteins nor enzymes. They include proteins that form receptors for certain substances on cell membranes and antibodies that participate in the body's immune responses (Chapter 17).

Enzymes

Enzymes increase the rate at which chemical reactions take place within living organisms in the temperature range compatible with life. We discuss enzymes in more detail in Chapter 5 but summarize their properties here. In general, enzymes speed up reactions by decreasing the energy required to start reactions. They also hold reactant molecules close together in the proper orientation for reactions to occur. Each enzyme has an **active site,** which is the site at which it combines with its **substrate,** the substance on which an enzyme acts. Enzymes have **specificity**—that is, each enzyme acts on a particular substrate or on a certain kind of chemical bond.

Like catalysts in inorganic chemical reactions, enzymes are not permanently affected or used up in the reactions they initiate. Enzyme molecules can be used over and over again to catalyze a reaction, although not indefinitely. Because enzymes are proteins, they are denatured by extremes of temperature and pH.

Nucleotides and Nucleic Acids

The chemical properties of *nucleotides* allow these compounds to perform several essential functions. One key function is storage of energy in **high-energy bonds**—bonds that, when broken, release more energy than do most covalent bonds. Nucleotides joined to form *nucleic acids* are, perhaps, the most remarkable of all biochemical substances. They store information that directs protein synthesis and that can be transferred from parent to progeny.

A **nucleotide** consists of three parts: (1) a nitrogenous base, so named because it contains nitrogen and has alkaline properties; (2) a five-carbon sugar; and (3) one or more phosphate groups, as ➢ Figure 2.20a shows for the nucleotide *adenosine triphosphate (ATP)*. The sugar and base alone make up a *nucleoside* (➢ Figure 2.20b).

The nucleotide ATP is the main source of energy in cells because it stores chemical energy in a form cells can use. The bonds between phosphates in ATP that are high-energy bonds are designated by wavy lines (➢ Figure 2.20c). They contain more energy than most covalent bonds, in that more energy is released when they are broken. Enzymes control the forming and breaking of high-energy bonds so that energy is released as needed within cells. The capture, storage, and use of energy is an important component of cellular metabolism (Chapter 5).

Nucleic acids consist of long polymers of nucleotides, called **polynucleotides.** They contain genetic information that determines all the heritable characteristics of a living organism, be it a microbe or a human. Such information is passed from generation to generation and di-

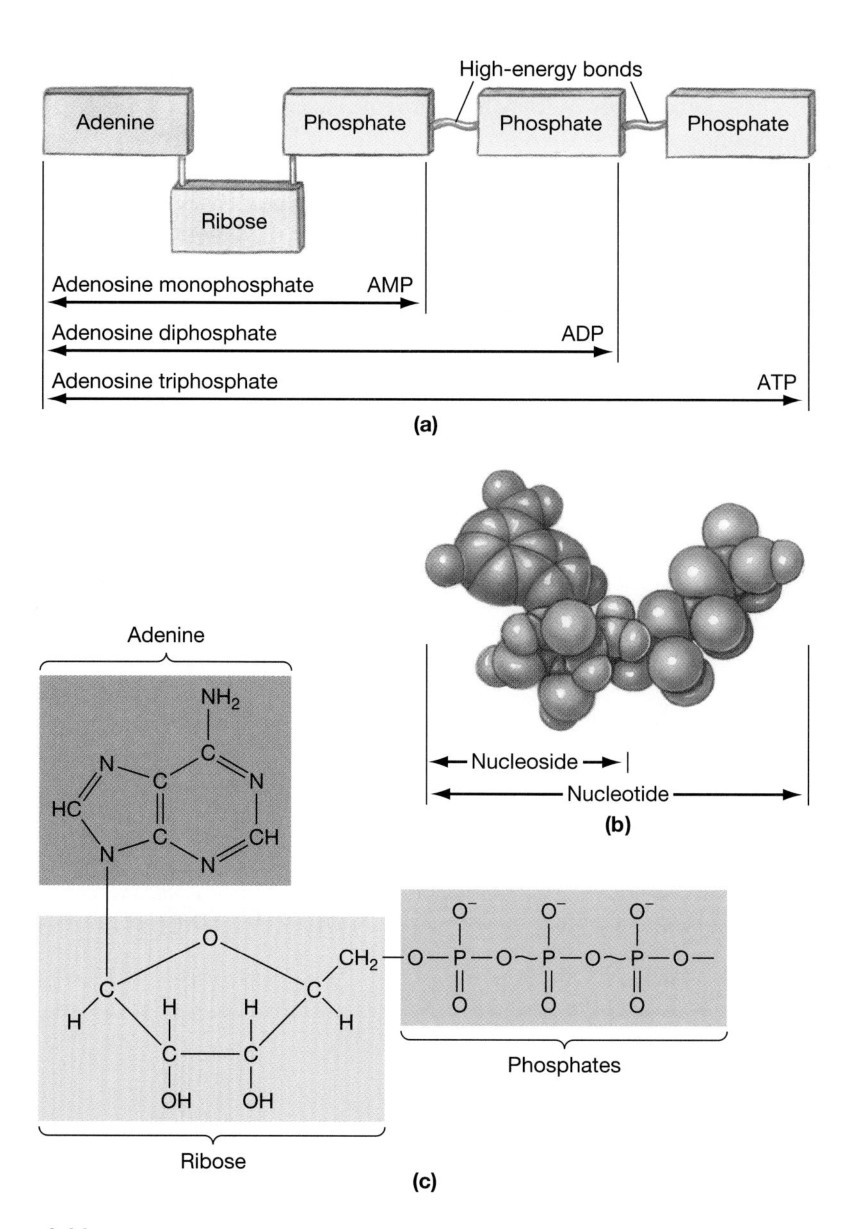

➢ **Figure 2.20** **Nucleotides.** (a) A nucleotide consists of a nitrogenous base, a five-carbon sugar, and one or more phosphate groups. (b) A nucleoside is comprised of the sugar and base without the phosphates. (c) The nucleotide adenosine triphosphate (ATP), the immediate source of energy for most activities of living cells. In ATP the base is adenine, and the sugar is ribose. Adding a phosphate group to adenosine diphosphate greatly increases the energy of the molecule; removal of the third phosphate group releases energy that can be used by the cell.

rects protein synthesis in each organism. By directing protein synthesis, nucleic acids determine which structural proteins and enzymes an organism will have. The enzymes determine which other substances the organism can make and which other reactions it can carry out.

The two nucleic acids found in living organisms are **ribonucleic acid (RNA)** and **deoxyribonucleic acid (DNA).** Except in a few viruses, RNA is a single polynucleotide chain, and DNA is a double chain of polynucleotides arranged as a double helix. In both nucleic acids, the phosphate and sugar molecules form a sturdy but inert "backbone" from which nitrogenous bases protrude. In DNA each chain is connected by hydrogen bonds between the bases, so the whole molecule resembles a ladder with many rungs (➢ Figure 2.21a).

DNA and RNA contain somewhat different building blocks (Table 2.5). RNA contains the sugar ribose, whereas DNA contains deoxyribose, which has one less oxygen atom than ribose. Three nitrogenous bases, adenine, cytosine, and guanine, are found in both DNA and RNA. In addition, DNA contains the base thymine, and RNA contains the base uracil. Of these bases, adenine and guanine are **purines,** nitrogenous base molecules that contain double-ring structures, and thymine, cytosine, and uracil are **pyrimidines,** nitrogenous base molecules that contain a single-ring structure (➢ Figure 2.22). All cellular organisms have both DNA and RNA. Viruses have either DNA or RNA but not both.

The two nucleotide chains of DNA are held together by hydrogen bonds between the bases and by other forces. The hydrogen bonds always connect adenine to thymine and cytosine to guanine, as shown in ➢ Figure 2.21b. This linking of specific bases is called **complemen-**

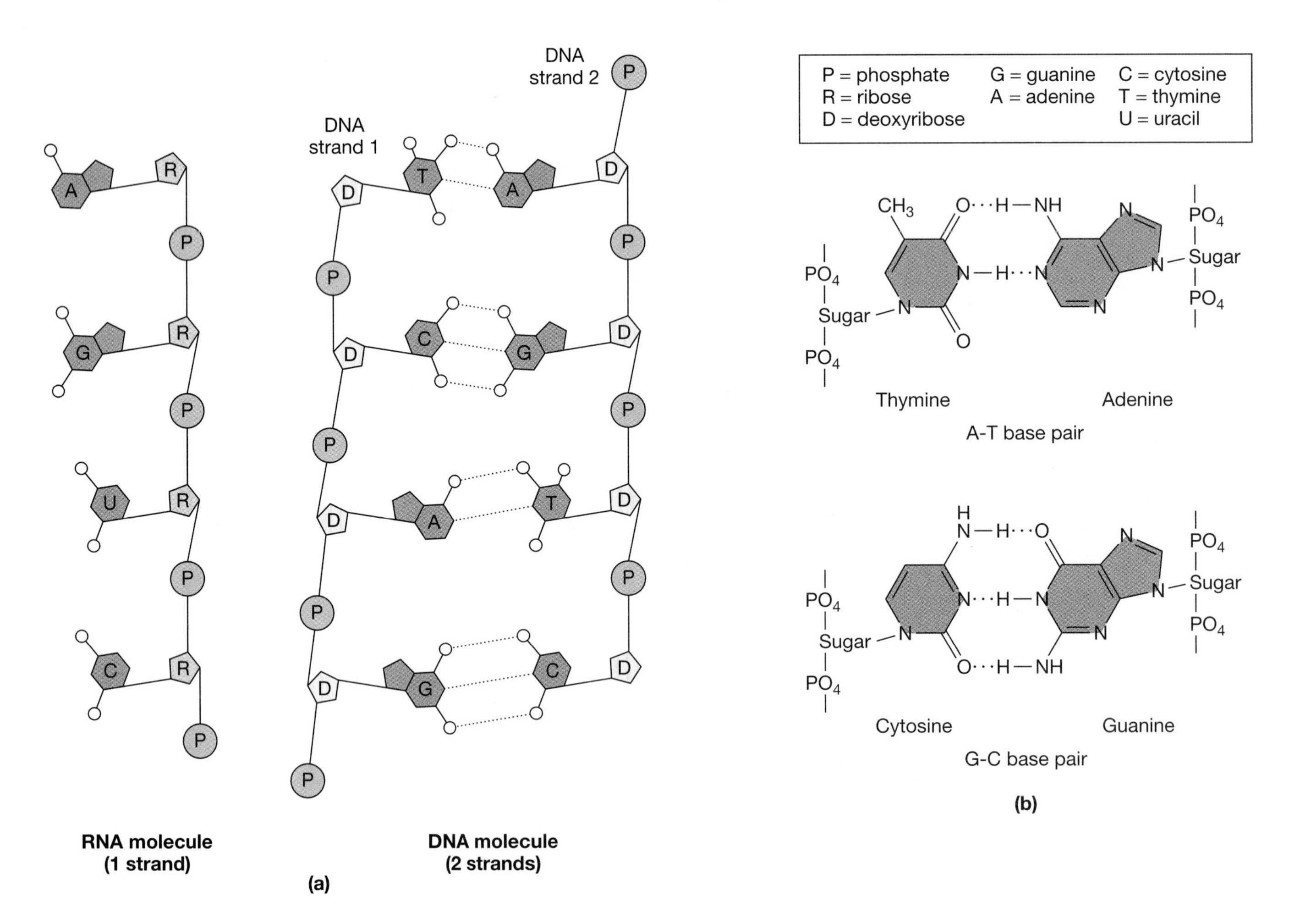

➢ Figure 2.21 **Nucleic and structure.** Nucleic acids consist of a backbone of alternating sugar and phosphate groups to which nitrogenous bases are attached. (a) RNA is usually single-stranded. DNA molecules typically consist of two chains held together by hydrogen bonds between bases. (b) The complementary base pairs in DNA, showing how hydrogen bonds are formed.

➢ Table 2.5 **Components of DNA and RNA**

Component		DNA	RNA
	Phosphoric acid	X	X
Sugars	Ribose		X
	Deoxyribose	X	
Bases	Adenine	X	X
	Guanine	X	X
	Cytosine	X	X
	Thymine	X	
	Uracil		X

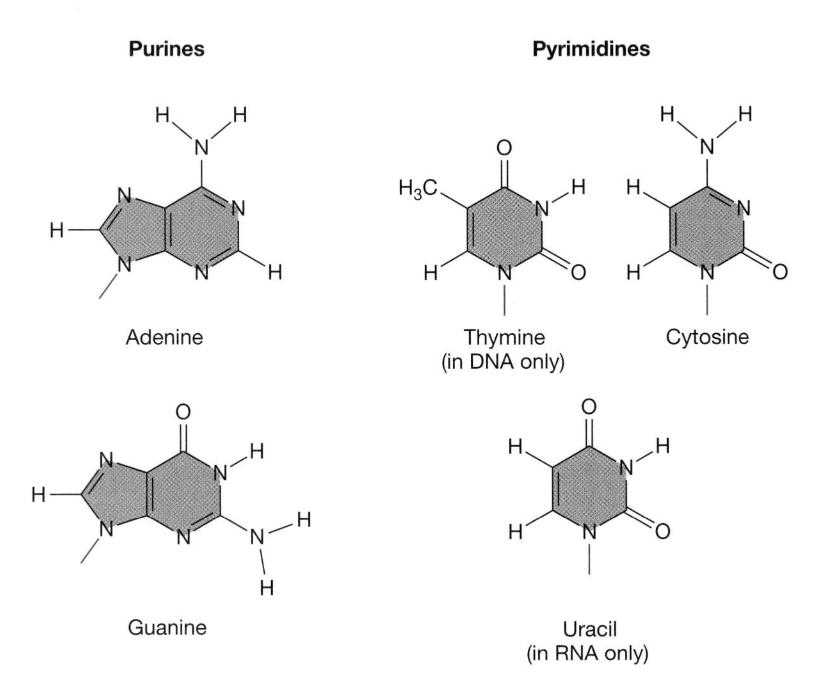

➢ Figure 2.22 **The five bases found in nucleic acids.** DNA contains the purines adenine and guanine and the pyrimidines cytosine and thymine. In RNA, thymine is replaced by the pyrimidine uracil.

tary base pairing. It is determined by the sizes and shapes of the bases. The same kind of complementary base pairing also occurs when information is transmitted from DNA to RNA at the beginning of protein synthesis (Chapter 7). In that situation, adenine in DNA base pairs with uracil in RNA.

DNA and RNA chains contain hundreds or thousands of nucleotides with bases arranged in a particular sequence. This sequence of nucleotides, like the sequence of letters in words and sentences, contains information that determines what proteins an organism will have. As noted earlier, an organism's structural proteins and enzymes, in turn, determine what the organism is and what it can do. Like changing a letter in a word, changing a nucleotide in a sequence can change the information it carries. The number of different possible sequences of bases is almost infinite, so DNA and RNA can contain a great many different pieces of information.

The functions of DNA and RNA are related to their ability to convey information. DNA is transmitted from one generation to the next. It determines the heritable characteristics of the new individual by supplying the information for the proteins its cells will contain. In contrast, RNA carries information from the DNA to the sites where proteins are manufactured in cells. There it directs and participates in the actual assembly of proteins. The functions of nucleic acids are discussed in more detail in Chapters 7 and 8.

✓ Why are starch, DNA, and RNA all considered to be polymers?

✓ Distinguish among primary, secondary, tertiary, and quaternary levels of protein structure.

Retracing Our Steps

Why Study Chemistry?

- A knowledge of basic chemistry is needed to understand how microorganisms function and how they affect humans and our environment.

Chemical Building Blocks and Chemical Bonds

Chemical Building Blocks

- The smallest chemical unit of matter is an **atom.** An **element** is a fundamental kind of matter, and the smallest unit of an element is an atom. An element is composed of only one type of atom. A **molecule** consists of two or more atoms chemically combined, and a **compound** consists of two or more different kinds of atoms chemically combined.
- The most common elements in all forms of life are carbon (C), hydrogen (H), oxygen (O), and nitrogen (N).

The Structure of Atoms

- Atoms consist of positively charged **protons** and neutral **neutrons** in the atomic nucleus and very small, negatively charged **electrons** orbiting the nucleus.
- The number of protons in an atom is equal to its **atomic number**. The total number of protons and neutrons determines the element's **atomic weight.**
- **Ions** are atoms that have gained or lost one or more electrons.
- **Isotopes** are atoms of the same element that contain different numbers of neutrons; some may be **radioisotopes.**

Chemical Bonds

- Atoms of molecules are held together by **chemical bonds.**
- **Ionic bonds** involve attraction of oppositely charged ions. In **covalent bonds,** atoms share pairs of electrons. **Hydrogen bonds** are weak attractions between polar regions of hydrogen atoms and oxygen or nitrogen atoms.

Chemical Reactions

- Chemical reactions involve breaking or forming chemical bonds and associated energy changes.
- **Catabolism,** the breaking down of molecules, releases energy. **Anabolism,** the synthesis of larger molecules, requires energy.

Water and Solutions

Water

- Water is a **polar compound,** acts as a solvent, and forms thin layers because it has high **surface tension.**
- Water also has high specific heat, and it serves as a medium for and participates in many chemical reactions.

Solutions and Colloids

- **Solutions** consist of **mixtures** with one or more **solutes** evenly distributed throughout a **solvent.**
- **Colloids** contain particles too large to form true solutions.

Acids, Bases, and pH

- In most solutions containing acids or bases, **acids** release H^+ ions, and **bases** accept H^+ ions (or release OH^- ions).
- The **pH** of a solution is a measure of its acidity or alkalinity. A pH of 7 is neutral, below 7 is acidic, and above 7 is basic, or **alkaline.**

Complex Organic Molecules

- **Organic chemistry** is the study of carbon-containing compounds.
- Organic compounds such as alcohols, aldehydes, ketones, organic acids, and amino acids can be identified by their **functional groups.**

Carbohydrates

- **Carbohydrates** consist of carbon chains in which most of the carbon atoms have an associated alcohol group and one carbon has either an aldehyde or a ketone group.
- The simplest carbohydrates are **monosaccharides,** which can combine to form **disaccharides** and **polysaccharides.** Long chains of repeating units are called **polymers.**
- The body uses carbohydrates primarily for energy.

Lipids

- All **lipids** are insoluble in water but soluble in nonpolar solvents.
- **Fats** consist of glycerol and **fatty acids.**
- **Phospholipids** contain a phosphate group in place of a fatty acid.
- **Steroids** have a complex four-ring structure.

Proteins

- **Proteins** consist of chains of **amino acids** linked by **peptide bonds.**
- Proteins form part of the structure of cells, act as enzymes, and contribute to other functions such as motility, transport, and regulation.
- **Enzymes** are biological catalysts of great **specificity** that increase the rate of chemical reactions in living organisms. Each enzyme has an **active site** to which its **substrate** binds.

Nucleotides and Nucleic Acids

- A **nucleotide** consists of a nitrogenous base, a sugar, and one or more phosphates.
- Some nucleotides contain **high-energy bonds.**
- **Nucleic acids** are important information-containing molecules that consist of chains of nucleotides. The nucleic acids that occur in living organisms are **ribonucleic acid (RNA)** and **deoxyribonucleic acid (DNA).**

Terminology Check

acid (*p. 337*)
active site (*p. 42*)
alkaline (*p. 34*)
amino acid (*p. 41*)
anabolism (*p. 30*)
anion (*p. 28*)
atom (*p. 26*)
atomic number (*p. 27*)
atomic weight (*p. 28*)
base (*p. 33*)
biochemistry (*p. 34*)
carbohydrate (*p. 35*)
catabolism (*p. 30*)
cation (*p. 28*)
chemical bond (*p. 29*)
colloid (*p. 32*)
complementary base pairing (*p. 44*)
compound (*p. 27*)
covalent bond (*p. 29*)
dehydration synthesis (*p. 32*)
denaturation (*p. 41*)
deoxyribonucleic acid (DNA) (*p. 44*)
disaccharide (*p. 36*)
electron (*p. 27*)
element (*p. 26*)
endergonic (*p. 31*)
enzyme (*p. 42*)
exergonic (*p. 30*)
fat (*p. 38*)
fatty acid (*p. 38*)
functional group (*p. 34*)
glycosidic bond (*p. 36*)
gram molecular weight (*p. 28*)
high-energy bond (*p. 42*)
hydrogen bond (*p. 30*)
hydrolysis (*p. 32*)
ion (*p. 28*)
ionic bond (*p. 29*)
isomer (*p. 35*)
isotope (*p. 28*)
lipid (*p. 38*)
mixture (*p. 32*)
mole (*p. 28*)
molecule (*p. 27*)
monosaccharide (*p. 35*)
neutral (*p. 34*)
neutron (*p. 27*)
nucleic acid (*p. 43*)
nucleotide (*p. 43*)
organic chemistry (*p. 34*)
peptide bond (*p. 41*)
pH (*p. 34*)
phospholipid (*p. 38*)
polar compound (*p. 30*)
polymer (*p. 36*)
polynucleotide (*p. 42*)
polypeptide (*p. 41*)
polysaccharide (*p. 36*)
primary structure (*p. 41*)
protein (*p. 41*)
proton (*p. 27*)
purine (*p. 44*)
pyrimidine (*p. 44*)
quaternary structure (*p. 41*)
radioisotope (*p. 29*)
reactant (*p. 32*)
R group (*p. 41*)
ribonucleic acid (RNA) (*p. 44*)
rule of octets (*p. 28*)
saturated fatty acid (*p. 38*)
secondary structure (*p. 41*)
solute (*p. 32*)
solution (*p. 32*)
solvent (*p. 32*)
specificity (*p. 42*)
steroid (*p. 39*)
structural protein (*p. 41*)
substrate (*p. 42*)
surface tension (*p. 31*)
tertiary structure (*p. 41*)
triacylglycerol (*p. 38*)
unsaturated fatty acid (*p. 38*)

See You on the Web

Don't forget I am expecting you on the web. The following is a preparedness checklist for your web explorations. If you have questions on any of these items, you can return to the end of Chapter 1 for more details.

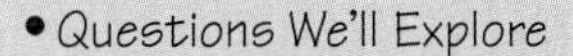

- Questions We'll Explore
- Concept Check questions
- Flashcards
- Study Guide

If you are ready, come visit me at

http://www.wiley.com/college/black

It will be a journey you'll never forget.

Special Staining Procedures

Negative Staining **Negative stains** are used when a specimen—or a part of it, such as the capsule—resists taking up a stain. The *capsule* is a layer of polysaccharide material that surrounds many bacterial cells and can act as a barrier to host defense mechanisms. It also repels stains. In negative staining, the background around the organisms is filled with a stain, such as India ink, or an acidic dye, such as nigrosin. This process leaves the organisms themselves as clear, unstained objects that stand out against the dark background. A second simple or differential stain can be used to demonstrate the presence of the cell inside the capsule. Thus, a typical slide will show a dark background and clear, unstained areas of capsular material, inside of which are purple cells stained with crystal violet (➢ Figure 3.30) or blue cells stained with methylene blue.

Flagellar Staining *Flagella,* appendages that some cells have and use for locomotion, are too thin to be seen easily with the light microscope. When it is necessary to determine their presence or arrangement, **flagellar stains** are painstakingly prepared to coat the surfaces of the flagella with dye or a metal such as silver. These techniques are very difficult and time-consuming and so are usually omitted from the beginning course in microbiology. (See Figure 4.12 for some examples of stained flagella.)

Endospore Staining A few types of bacteria produce resistant cells called *endospores*. Endospore walls are very resistant to penetration of ordinary stains. When a simple stain is used, the spores will be seen as clear, glassy, easily recognizable areas within the bacterial cell. Thus, strictly speaking, it is not absolutely necessary to perform an endospore stain to see the spores. However, the differential **Schaeffer-Fulton spore stain** makes spores easier to visualize (➢ Figure 3.31). Heat-fixed smears are covered with malachite green and then gently heated until they steam. Approximately 5 minutes of such steaming causes the endospore walls to become more permeable to the dye. The slide is then rinsed with water for 30 seconds to remove the green dye from all parts of the cell except for the endospores, which retain it. Then a counterstain of safranin is placed on the slide to stain the non-spore-forming, or vegetative, areas of the cells. Cells of cultures without endospores appear red; those with endospores have green spores and red vegetative cells.

Although microscopy and staining techniques can offer valuable information about microorganisms, these methods are usually not enough to permit identification of most microbes. Many species look identical under the microscope—after all, there are only a limited number of basic shapes, arrangements, and staining reactions but thousands of kinds of bacteria. This means that biochemical and genetic characteristics usually must be determined before an identification can be made (Chapter 9).

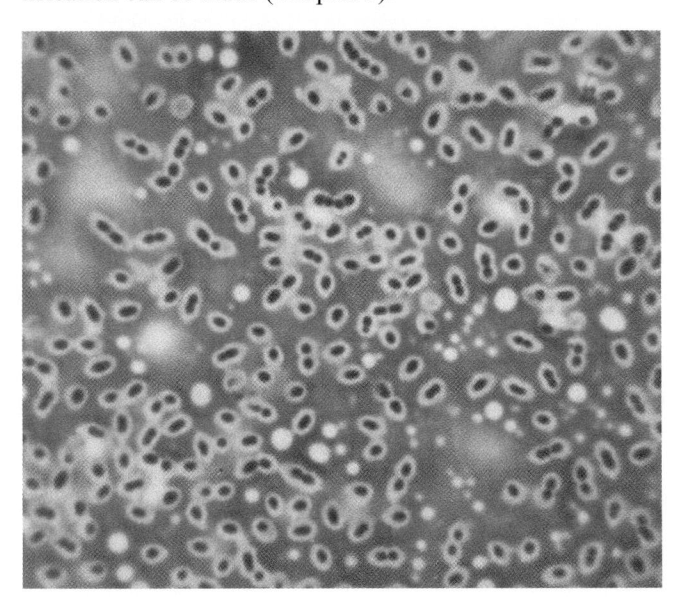

➢ Figure 3.30 **Negative staining.** Negative staining for capsules reveals a clear area (the capsule, which does not accept stain) in a dark pink background of India ink and crystal violet counterstain. The cells themselves are stained deep purple with the counterstain. The bacteria are *Streptococcus pneumoniae* (3300X), which are arranged in pairs.

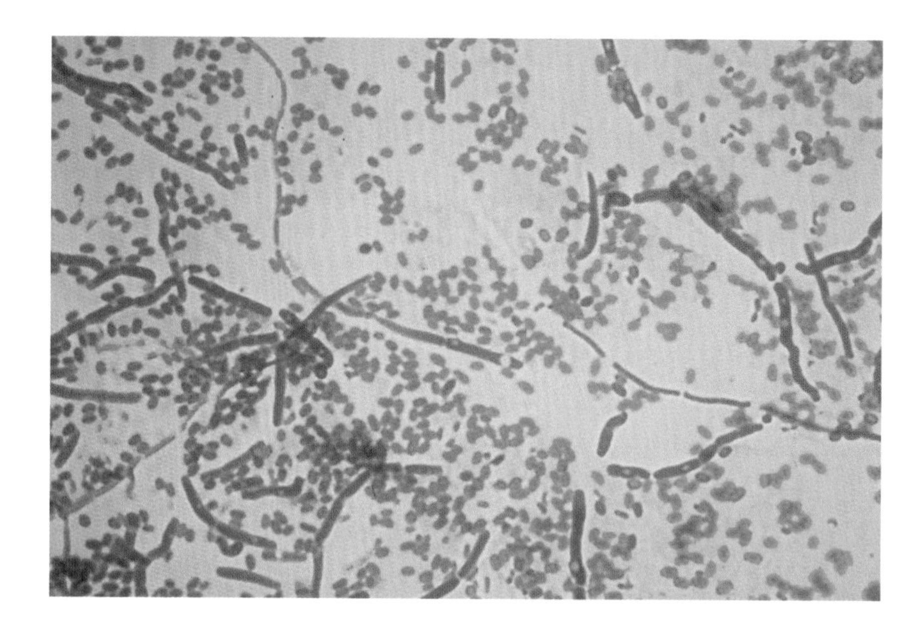

➢ Figure 3.31 **The Schaeffer-Fulton spore stain.** Endospores of *Bacillus megaterium* (1000X) are visible as green, oval structures inside and outside the rod-shaped cells. Vegetative cells, which represent a non-spore-forming stage, and cellular regions without spores stain red.

Retracing Our Steps

Historical Microscopy

- The existence of microorganisms was unknown until the invention of the microscope. Leeuwenhoek, probably the first to see microorganisms (in the 1600s), set the stage for **microscopy,** the technology of making very small things visible to the human eye.
- Leeuwenhoek's simple microscopes could reveal little detail of specimens. Today, multiple-lens, compound microscopes give us nearly distortion-free images, enabling us to delve further into the study of microbes.

Principles of Microscopy

Metric Units

- The three units most used to describe microbes are the **micrometer** (μm), formerly called a micron, which is equal to 0.000001 m, also written as 10^{-6} m; the **nanometer** (nm), formerly called a millimicron (mμ), which is equal to 0.000000001 m, or 10^{-9} m; and the **angstrom** (Å), which is equal to 0.0000000001 m, 0.1 nm, or 10^{-10} m, but is no longer officially recognized.

Properties of Light: Wavelength and Resolution

- The **wavelength,** or the length of light rays, is the limiting factor in resolution.
- **Resolution** is the ability to see two objects as separate, discrete entities. Light wavelengths must be small enough to fit between two objects in order for them to be resolved.
- **Resolving power** can be defined as RP = $\lambda/2\text{NA}$, where λ = wavelength of light. The smaller the value of λ and the larger the value of NA, the greater the resolving power of the lens.
- **Numerical aperture** (NA) relates to the extent to which light is concentrated by the condenser and collected by the objective. Its value is engraved on the side of each objective lens.

Properties of Light: Light and Objects

- If light strikes an object and bounces back, **reflection** (which gives an object its color) has occurred.
- **Transmission** is the passage of light through an object. Light must either be reflected from or transmitted through an object for it to be seen with a light microscope.
- **Absorption** of light rays occurs when they neither bounce off nor pass through an object but are taken up by that object. Absorbed light energy is used in performing photosynthesis or in raising the temperature of the irradiated body.
- Reemission of absorbed light as light of longer wavelengths is known as **luminescence.** If reemission occurs only during irradiation, the object is said to **fluoresce.** If reemission continues after irradiation ceases, the object is said to be **phosphorescent.**
- **Refraction** is the bending of light as it passes from one medium to another of different density. **Immersion oil,** which has the same **index of refraction** as glass, is used to replace air and to prevent refraction at a glass-air interface.
- **Diffraction** occurs when light waves are bent as they pass through a small opening, such as a hole, a slit, a space between two adjacent cellular structures, or a small, high-powered, magnifying lens in a microscope. The bent light rays distort the image obtained and limit the usefulness of the light microscope.

Light Microscopy

The Compound Light Microscope

The major parts of a compound light microscope and their functions are as follows:

- **Base** Supporting structure that generally contains the light source.
- **Condenser** Converges light beams to pass through the specimen.

- **Iris diaphragm** Controls the amount of light passing through the specimen.
- **Objective lens** Magnifies image.
- **Body tube** Conveys light to the ocular lens.
- **Ocular lens** Magnifies the image from the objective. A microscope with one ocular lens (eyepiece) is **monocular;** a microscope with two oculars is **binocular.**
- **Mechanical stage** Allows precise control in moving the slide.
- **Coarse adjustment** Knob used to locate specimen.
- **Fine adjustment** Knob used to bring specimen into sharp focus.
- The **total magnification** of a light microscope is calculated by multiplying the magnifying power of the objective lens by the magnifying power of the ocular lens. Increased magnification is of no value unless good resolution can also be maintained.

Dark-Field Microscopy

- **Bright-field illumination** is used in the ordinary light microscope, with light passing directly through the specimen.
- **Dark-field illumination** utilizes a special condenser that causes light to reflect off the specimen at an angle rather than pass directly through it.

Phase-Contrast Microscopy

- **Phase-contrast microscopy** uses microscopes with special condensers that accentuate small differences in refractive index of structures within the cell, allowing live, unstained organisms to be examined.

Nomarski (Differential Interference Contrast) Microscopy

- **Nomarski microscopy** uses microscopes that operate essentially like phase-contrast microscopes but with a much greater resolution and a very short depth of field. They produce a nearly three-dimensional image.

Fluorescence Microscopy

- **Fluorescence microscopy** uses ultraviolet light instead of white light to excite molecules within the specimen or dye molecules attached to the specimen. These molecules emit different wavelengths, often of brilliant colors.

Electron Microscopy

- The **electron microscope (EM)** uses a beam of electrons instead of a beam of light and electromagnets instead of glass lenses for focusing. They are much more expensive and difficult to use but give magnifications of up to 500,000X and a resolving power of less than 1 nm. Viruses can be seen only by using EMs.
- Advanced types of electron microscopes can visualize actual molecules and individual atoms.

Transmission Electron Microscopy

- For the **transmission electron microscope (TEM)**, very thin slices (sections) of a specimen are used, revealing the internal structure of microbial and other cells.

Scanning Electron Microscopy

- For the **scanning electron microscope (SEM)**, a specimen is coated with a metal. The electron beam is scanned, or swept, over this coating to form a three-dimensional image.

Scanning Tunneling Microscopy

- The **scanning tunneling microscope (STM)** can produce three-dimensional images of individual molecules and atoms, as well as movies.

Techniques of Light Microscopy

Preparation of Specimens for the Light Microscope

- **Wet mounts** are used to view living organisms. The **hanging-drop** technique is a special type of wet mount, often used to determine whether organisms are motile.
- **Smears** of appropriate thickness are allowed to air-dry completely and are then passed through an open flame. This process, called **heat fixation,** kills the organisms, causing them to adhere to the slide and more readily accept stains.

Principles of Staining

- A **stain,** or dye, is a molecule that can bind to a structure and give it color.
- Most microbial stains are **cationic** (positively charged), or **basic, dyes,** such as methylene blue. Because most bacterial surfaces are negatively charged, these dyes are attracted to them.
- **Simple stains** use one dye and reveal basic cell shapes and arrangements. **Differential stains** use two or more dyes and distinguish various properties of organisms. The **Gram stain,** the **Schaeffer-Fulton spore stain,** and the **Ziehl-Neelsen acid-fast stain** are examples.
- **Negative stains** color the background around cells and their parts, which resist taking up stain.
- **Flagellar stains** add layers of dye or metal to the surfaces of flagella to make those surfaces visible.
- In the Schaeffer-Fulton spore stain, endospores stain green due to the uptake of malachite green, whereas vegetative cells stain red due to safranin uptake.

Terminology Check

absorption (*p. 53*)
angstrom (*p. 49*)
anionic (acidic) dye (*p. 64*)
atomic force microscope (AFM) (*p. 62*)
base (*p. 55*)
binocular (*p. 55*)
body tube (*p. 55*)
bright-field illumination (*p. 56*)
cationic (basic) dye (*p. 64*)
coarse adjustment (*p. 55*)
compound light microscope (*p. 54*)
condenser (*p. 55*)
dark-field illumination (*p. 56*)
differential stain (*p. 64*)
diffraction (*p. 54*)

electron micrograph (*p. 58*)
electron microscope (EM) (*p. 57*)
fine adjustment (*p. 55*)
flagellar stain (*p. 67*)
fluoresce (*p. 53*)
fluorescence microscopy (*p. 57*)
fluorescent antibody staining (*p. 57*)
freeze-etching (*p. 59*)
freeze-fracturing (*p. 59*)
Gram stain (*p. 64*)
hanging drop (*p. 62*)
heat fixation (*p. 62*)
immersion oil (*p. 53*)
index of refraction (*p. 53*)
iris diaphragm (*p. 55*)
light microscopy (*p. 54*)
luminescence (*p. 53*)
mechanical stage (*p. 55*)
micrometer (*p. 49*)
microscopy (*p. 49*)
monocular (*p. 55*)
mordant (*p. 64*)
nanometer (*p. 49*)
negative stain (*p. 67*)
Nomarski microscopy (*p. 57*)
numerical aperture (NA) (*p. 51*)
objective lens (*p. 55*)
ocular lens (*p. 55*)
ocular micrometer (*p. 55*)
optical microscope (*p. 54*)
parfocal (*p. 55*)
phase-contrast microscopy (*p. 57*)
phosphorescent (*p. 53*)
reflection (*p. 52*)
refraction (*p. 53*)
resolution (*p. 50*)
resolving power (RP) (*p. 51*)
scanning electron microscope (SEM) (*p. 60*)
scanning tunneling microscope (STM) (*p. 60*)
Schaeffer-Fulton spore stain (*p. 67*)
shadow casting (*p. 59*)
simple stain (*p. 64*)
smear (*p. 62*)
stain (*p. 63*)
total magnification (*p. 55*)
transmission (*p. 53*)
transmission electron microscope (TEM) (*p. 58*)
wavelength (*p. 50*)
wet mount (*p. 62*)
Ziehl-Neelsen acid-fast stain (*p. 65*)

See You on the Web

Don't forget I am expecting you on the web. The following is a preparedness checklist for your web explorations. If you have questions on any of these items, you can return to the end of Chapter 1 for more details.

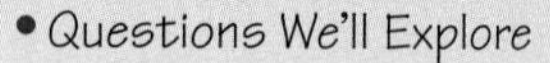

- Questions We'll Explore
- Concept Check questions
- Flashcards
- Study Guide

If you are ready, come visit me at

http://www.wiley.com/college/black

It will be a journey you'll never forget.

Characteristics of Prokaryotic and Eukaryotic Cells

Overwhelmed by all these doors? Which one do you open? Where will it take you? Will you be able to pass through it? Maybe they're all locked. What then?

Perhaps these are questions you'd ask if you were a relatively large molecule trying to get into or out of a cell. Just as there is a wonderful diversity of doors, there is also a diversity of protein transport molecules found in cell membranes. These protein transport molecules control what enters and leaves the cell. They can let some molecules in while keeping others out. They might require that one molecule go out at the same time as a different molecule is coming in. In their various forms, protein transport molecules help control traffic across the cell membrane.

Background Basics

Gram stain . . . . *(Chapter 3)*
acid-fast stain . . . . *(Chapter 3)*
polar bonding . . . . *(Chapter 2)*
covalent bonding . . . . *(Chapter 2)*
steroid . . . . *(Chapter 2)*
protein . . . . *(Chapter 2)*
ion . . . . *(Chapter 2)*
DNA . . . . *(Chapter 2)*

Questions We'll Explore

A What are the characteristics of prokaryotic and eukaryotic cells?

B How do prokaryotic cells differ in size, shape, and arrangement?

C How are structure and function related in bacterial cell walls and cell membranes?

D How are structure and function related in other bacterial components?

E How are structure and function related in eukaryotic plasma membranes?

F How are structure and function related in other eukaryotic components?

G How do passive transport processes function, and why are they important? How does active transport function, and why is it important?

H How do exocytosis and endocytosis occur, and why are they important?

We have considered the chemical principles that apply to cells and how to use microscopes and stains to observe cells. We can now look at the structure and function of the cells themselves.

Basic Cell Types

All living cells can be classified as prokaryotic, from the Greek words *pro* (before) and *karyon* (nucleus) or eukaryotic, from *eu* (true) and *karyon* (nucleus). **Prokaryotic** (pro-kar″e-ot′ik) **cells** lack a nucleus and other membrane-enclosed structures, whereas **eukaryotic** (u-kar″e-ot′ik) **cells** have such structures.

All prokaryotes are single-celled organisms, as exemplified by the bacteria. Most of this book will be devoted to the study of prokaryotes. Eukaryotes include all plants, animals, fungi, and protists (organisms such as *Amoeba, Paramecium,* and the malaria parasite). We will also spend some time studying eukaryotes, especially the fungi and various parasites, plus the interactions of eukaryotic cells and prokaryotes.

Prokaryotic and eukaryotic cells are *similar* in several ways. Both are surrounded by a *cell membrane,* or *plasma membrane*. Although some cells have structures that extend beyond this membrane or surround it, the membrane defines the boundaries of the living cell. Both prokaryotic and eukaryotic cells also encode genetic information in DNA molecules.

These two types of cells are *different* in other, important ways. In eukaryotic cells, DNA is in a nucleus surrounded by a *nuclear envelope,* but in prokaryotic cells, DNA is in a nuclear region not surrounded by a membrane. Eukaryotic cells also have a variety of internal structures called **organelles** (or-ga-nelz′), or "little organs," that are surrounded by one or more membranes. Prokaryotic cells generally lack organelles that are membrane-enclosed. We take advantage of some of the differences between eukaryotic human cells and prokaryotic bacterial cells when we try to control disease-causing bacteria without harming the human host.

In this chapter we examine the similarities and differences of prokaryotic cells and eukaryotic cells, as summarized in Table 4.1. (Refer to this table each time you learn about a new cellular structure.) Viruses do not fit in either category, as they are acellular. However, some viruses infect prokaryotic cells, while other viruses infect eukaryotes. Chapter 10 will examine viruses in detail.

Table 4.1 Similarities and differences between prokaryotic and eukaryotic cells

Characteristic	Prokaryotic cells	Eukaryotic cells
Genetic structures		
Genetic material (DNA)	Found in single chromosome	Typically found in paired chromosomes
Location of genetic information	Nuclear region (nucleoid)	Membrane-enclosed nucleus
Nucleolus	Absent	Present
Histones	Absent	Present
Extrachromosomal DNA	In plasmids	In organelles, such as mitochondria and chloroplasts, and in plasmids
Intracellular structures		
Mitotic spindle	Absent	Present during cell division
Plasma membrane	Fluid-mosaic structure lacking sterols	Fluid-mosaic structure containing sterols
Internal membranes	Only in photosynthetic organisms	Numerous membrane-enclosed organelles
Endoplasmic reticulum	Absent	Present
Respiratory enzymes	Cell membrane	Mitochondria
Chromatophores	Present in photosynthetic bacteria	Absent
Chloroplasts	Absent	Present in some
Golgi apparatus	Absent	Present
Lysosomes	Absent	Present
Peroxisomes	Absent	Present
Ribosomes	70S	80S in cytoplasm and on endoplasmic reticulum, 70S in organelles
Cytoskeleton	Absent	Present
Extracellular structures		
Cell wall	Peptidoglycan found on most cells	Cellulose, chitin, or both found on plant and fungal cells
External layer	Capsule or slime layer	Pellicle, test, or shell in certain protists
Flagella	When present, consist of fibrils of flagellin	When present, consist of complex membrane-enclosed structure with "9 + 2" microtubule arrangement
Cilia	Absent	Present as structures shorter than, but similar to, flagella in some eukaryotic cells
Pili	Present as attachment or conjugation pili in some prokaryotic cells	Absent
Reproductive process		
Cell division	Binary fission	Mitosis and/or meiosis
Sexual exchange of genetic material	Not part of reproduction	Meiosis
Sexual or asexual reproduction	Only asexual reproduction	Sexual or asexual reproduction

Prokaryotic Cells

Detailed studies of cells have revealed that prokaryotes differ enough to be split into two large groups called *domains.* A relatively new concept in classification, domain is the highest taxonomic category, higher even than kingdom. Three domains exist: two prokaryotic and one eukaryotic:

- Archaea (archaeobacteria) (from *archae,* ancient)
- Bacteria (eubacteria)
- Eukarya

All members of Archaea and Bacteria are prokaryotes and have traditionally been called types of bacteria. A problem of terminology arises over the use of a capital versus a lowercase b in the word *bacteria.* All bacteria (lowercase b) are prokaryotes, but not all prokaryotes belong to the domain Bacteria (capital B). The differences between Archaea and Bacteria are not so much structural as molecular. Therefore, most of what we have to say about "bacteria" in this chapter applies to both Archaea and Bacteria. (We will discuss Archaea further in Chapter 11.

Most bacteria on this planet, both in the environment and living in and on humans, are members of the domain Bacteria. As yet, we know of no disease-causing Archaea. However, they are very important in the ecology of our planet, especially in extreme environments, such as in deep-sea hydrothermal vents where sulfur-laden water, at temperatures exceeding the boiling point of water, gushes out from openings in the ocean floor.

How Is a Bacterium Like a Lunch Meat?

Bacterial cell walls, organelles, and eukaryotic cells are diagrammed in your textbook to demonstrate their three-dimensional arrangements. Are these illustrations a figment of the imagination, or is there some way to construct complex three-dimensional structures of extremely small submacroscopic objects? Using serial sections (i.e., examining adjacent slices, one right after the next) is one method used to elucidate the three-dimensional relationships between microscopic cell parts. The determination of complex shapes using serial sections is not unlike determining the shape and size of the fat globules and peppercorns that lace a good salami. After making many orderly slices through the salami, the geometry of the salami inclusions is revealed as the shape from each section is superimposed over the preceding section. Using serial sections, in conjunction with chemical and biophysical data, makes it possible to produce the three-dimensional cell structures that illustrate your text.

Size, Shape, and Arrangement

Size

Prokaryotes are among the smallest of all organisms. Most prokaryotes range from 0.5 to 2.0 μm in diameter. For comparison, a human red blood cell is about 7.5 μm in diameter. Keep in mind, however, that although we often use diameter to specify cell size, many cells are not spherical in shape. Some spiral bacteria have a much larger diameter, and some cyanobacteria are 60 μm long. Because of their small size, bacteria have a large surface-to-volume ratio. For example, spherical bacteria with a diameter of 2 μm have a surface area of about 12 μm^2 and a volume of about 4 μm^3. Their surface-to-volume ratio is 12:4, or 3:1. In contrast, eukaryotic cells with a diameter of 20 μm have a surface area of about 1200 μm^2 and a volume of about 4000 μm^3. Their surface-to-volume ratio is 1200:4000, or 0.3:1—only one-tenth as great. The large surface-to-volume ratio of bacteria means that no internal part of the cell is very far from the surface and that nutrients can easily and quickly reach all parts of the cell.

Shape

Typically, bacteria display three basic shapes—spherical, rodlike, and spiral (➢ Figure 4.1)—but variations abound. A spherical bacterium is called a **coccus** (kok'us; plural: *cocci* [kok'se]), and a rodlike bacterium is called a **bacillus** (ba-sil'us; plural: *bacilli* [bas-il'e]). Some bacteria, called *coccobacilli,* are short rods intermediate in shape between cocci and bacilli. Spiral bacteria have a variety of curved shapes. A comma-shaped bacterium is called a **vibrio** (vib're-o); a rigid, wavy-shaped one, a **spirillum** (spi-ril'um; plural: *spirilla*); and a corkscrew-shaped one, a **spirochete** (spi'ro-ket). Some bacteria do not fit any of the preceding categories but rather have spindle shapes or irregular, lobed shapes. Square bacteria were discovered on the shores of the Red Sea in 1981. They are 2 to 4 μm on a side and sometimes aggregate in wafflelike sheets.

Even bacteria of the same kind sometimes vary in size and shape. When nutrients are abundant in the environment and cell division is rapid, rods are often twice as large as those in an environment with only a moderate supply of nutrients. Although variations in shape within a single species of bacteria are generally small, there are exceptions. Some bacteria vary widely in form even within a single culture, a phenomenon known as **pleomorphism.** Moreover, in aging cultures where organisms have used up most of the nutrients and have deposited wastes, cells not only are generally smaller, but they often display a great diversity of unusual shapes.

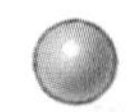

➢ Figure 4.1 **The most common bacterial shapes.**

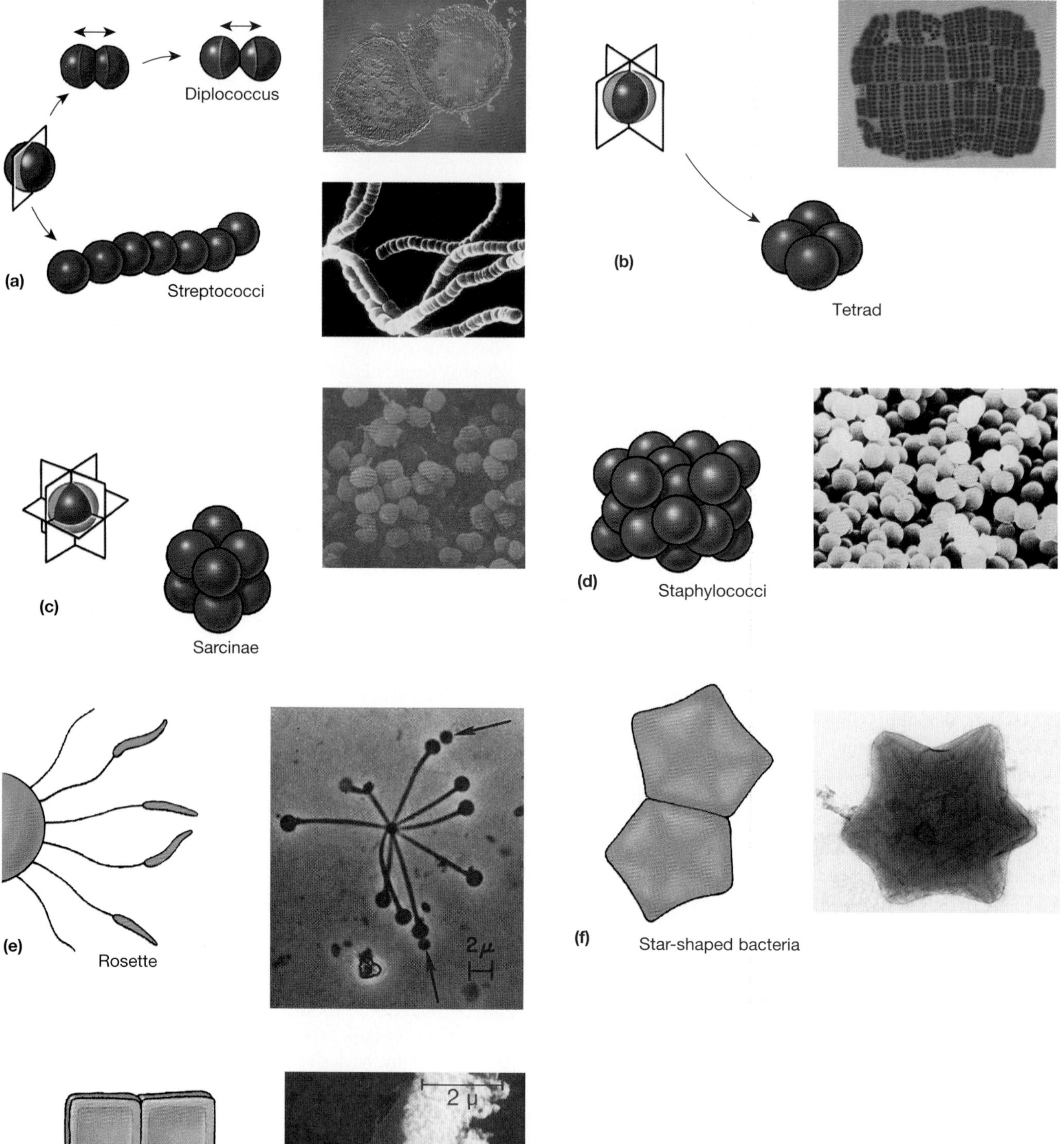

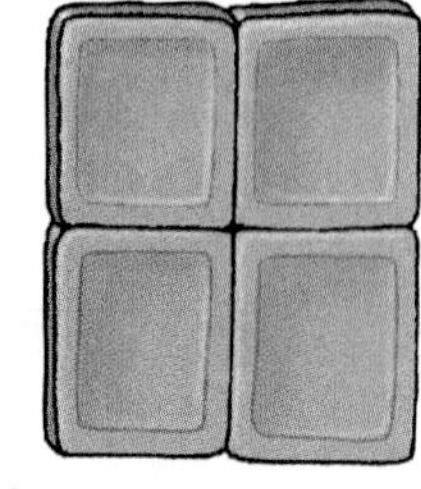

➤ Figure 4.2 **Arrangements of bacteria.** (a) Cocci arranged in pairs (diplococci of *Neisseria*) and in chains (*Streptococcus*), formed by division in one plane. (b) Cocci arranged in a tetrad (*Merisopedia,* 100X), formed by division in two planes. (c) Cocci arranged in a sarcina (*Sarcina lutea,* 16,000X), formed by division in three planes. (d) Cocci arranged in a cluster (*Staphylococcus,* 5400X), formed by division in many planes. (e) Bacilli arranged in a rosette (*Caulobacter,* 2400X), attached by stalks to a substrate. (f) Star-shaped bacteria (*Stella*). (g) Square-shaped bacteria.

A few substances also diffuse through pores. Such diffusion is affected by the size and charge of the diffusing particles and the charges on the pore surface. Pores probably have a diameter of less than 0.8 nm, so only water, small water-soluble molecules, and ions such as H^+, K^+, Na^+, and Cl^- pass through them. This is one reason a membrane is said to be **selectively permeable** (*semipermeable*).

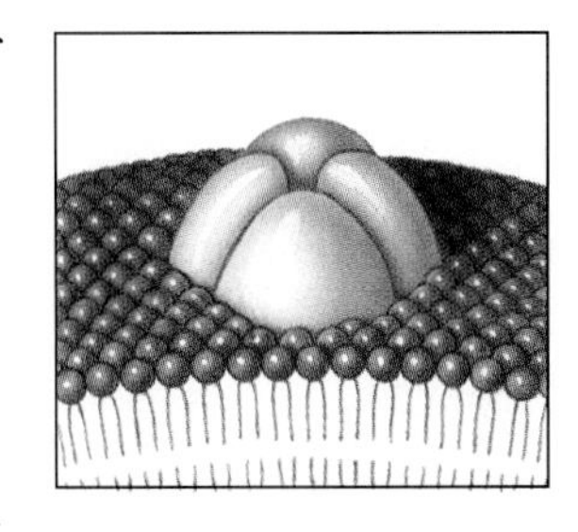

Facilitated Diffusion

Facilitated diffusion is diffusion down a concentration gradient and across a membrane with the assistance of special pores or carrier molecules. In fact, membranes contain protein-lined pores for specific ions. These pores have an arrangement of charges that allows rapid passage of a particular ion. The carrier molecules are proteins, embedded in the membrane, that bind to one or to a few specific molecules and assist in their movement. By one possible mechanism for facilitated diffusion, a carrier acts like a revolving door or shuttle that provides a convenient one-way channel for the movement of substances across a membrane (➢ Figure 4.29). Carrier molecules can become saturated, and similar molecules sometimes compete for the same carrier. Saturation occurs when all the carrier molecules are moving the diffusing substance as fast as they can. Under these conditions the rate of diffusion reaches a maximum and cannot increase further. When a carrier molecule can transport more than one substance, the substances compete for the carrier in proportion to their concentrations. For example, if there is twice as much of substance A as substance B, substance A will move across the membrane twice as fast as substance B.

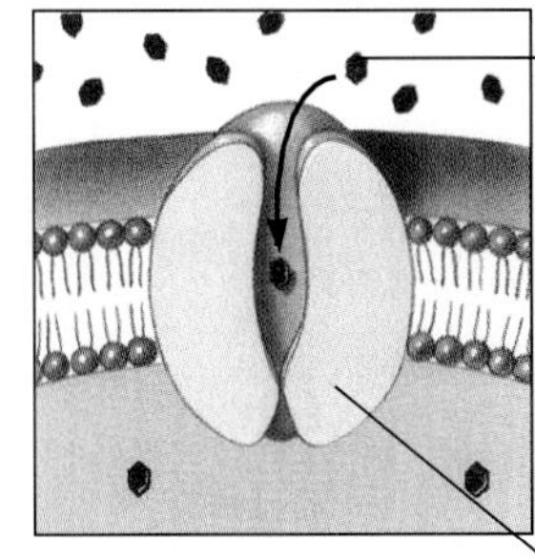

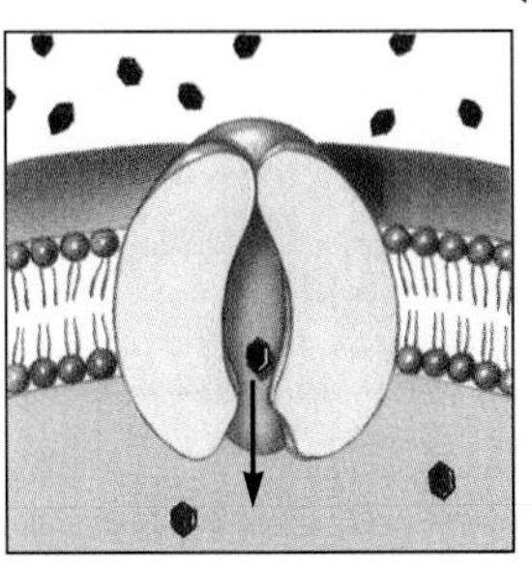

➢ Figure 4.29 **Facilitated diffusion.** Carrier protein molecules aid in the movement of substances through the cell membrane, but only down their concentration gradient (from a region where their concentration is high to one where their concentration is low). This process does not require the expenditure of any energy (ATP) by the cell.

Osmosis

Osmosis is a special case of diffusion-in which water molecules diffuse across a selectively permeable membrane. To demonstrate osmosis, we start with two compartments separated by a membrane permeable only to water. One compartment contains pure water, and the other compartment contains some large, nondiffusible molecules, such as proteins (➢ Figure 4.30a). Water molecules move in both directions, but their net movement is from pure water (concen-

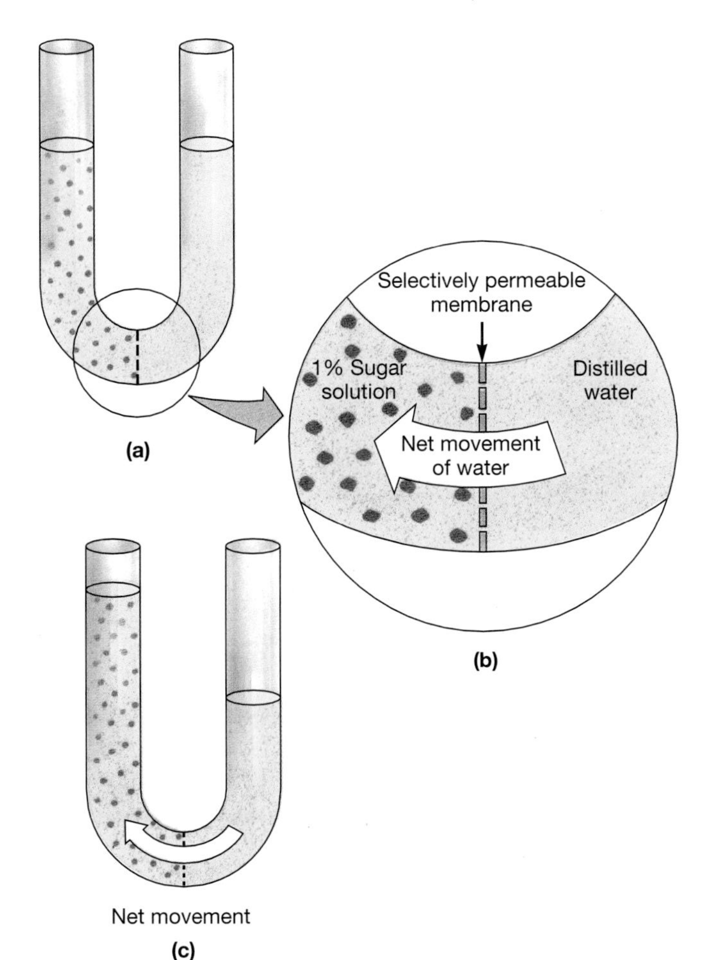

➢ Figure 4.30 **Osmosis.** (a) The diffusion of water from an area of higher water concentration (the right side) to an area of lower water concentration (the left side) through a semipermeable membrane. (b) Here the net movement of water is into the sugar solution because the concentration of water there is slightly lower than on the other side of the membrane. (c) As a result of the net movement of water, the column rises on the left.

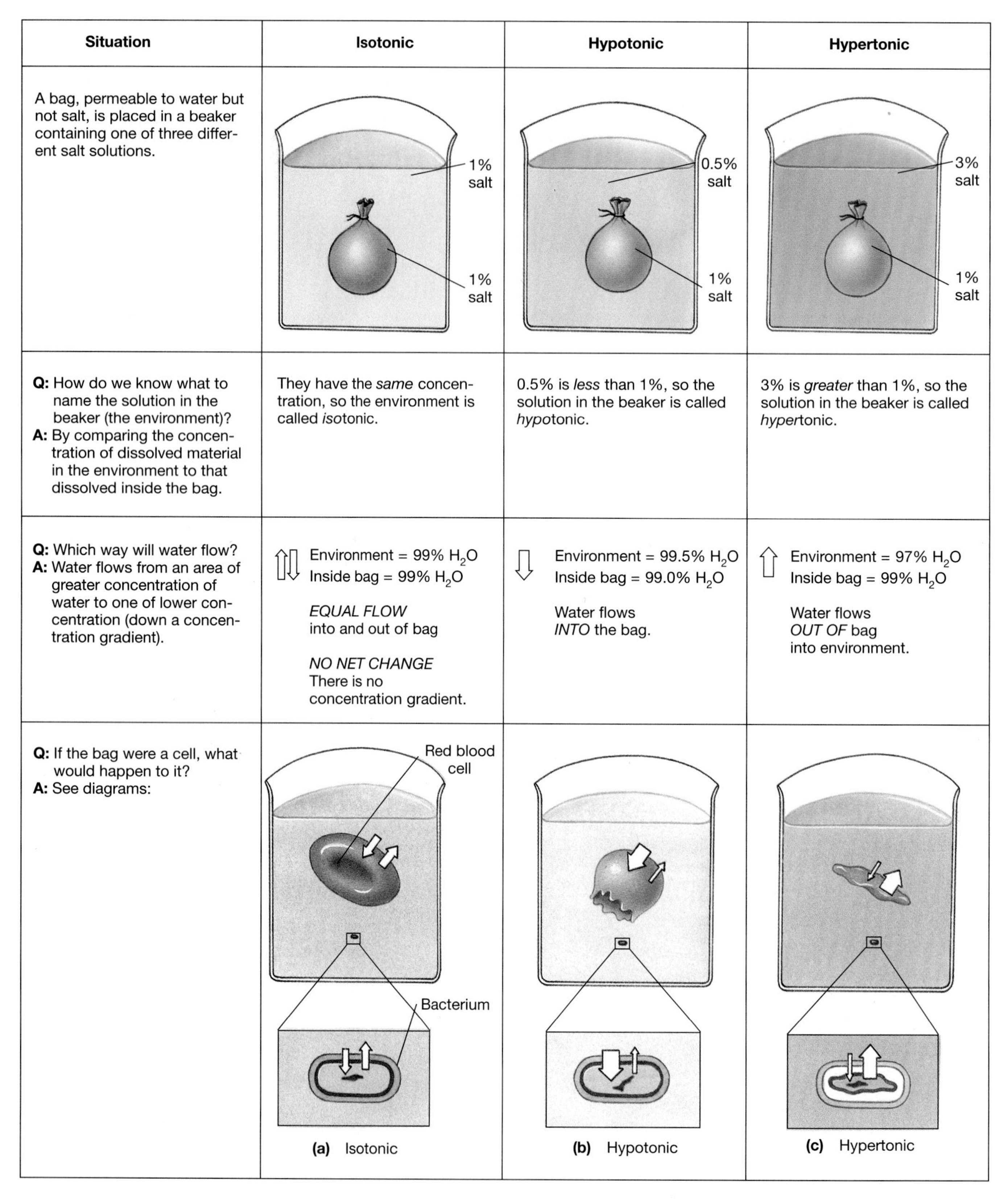

Situation	Isotonic	Hypotonic	Hypertonic
A bag, permeable to water but not salt, is placed in a beaker containing one of three different salt solutions.	1% salt 1% salt	0.5% salt 1% salt	3% salt 1% salt
Q: How do we know what to name the solution in the beaker (the environment)? **A:** By comparing the concentration of dissolved material in the environment to that dissolved inside the bag.	They have the *same* concentration, so the environment is called *iso*tonic.	0.5% is *less* than 1%, so the solution in the beaker is called *hypo*tonic.	3% is *greater* than 1%, so the solution in the beaker is called *hyper*tonic.
Q: Which way will water flow? **A:** Water flows from an area of greater concentration of water to one of lower concentration (down a concentration gradient).	Environment = 99% H_2O Inside bag = 99% H_2O *EQUAL FLOW* into and out of bag *NO NET CHANGE* There is no concentration gradient.	Environment = 99.5% H_2O Inside bag = 99.0% H_2O Water flows *INTO* the bag.	Environment = 97% H_2O Inside bag = 99% H_2O Water flows *OUT OF* bag into environment.
Q: If the bag were a cell, what would happen to it? **A:** See diagrams:	Red blood cell Bacterium **(a)** Isotonic	**(b)** Hypotonic	**(c)** Hypertonic

➢ Figure 4.31 **Experiments that examine the effects of tonicity on osmosis.** (a) A cell in an isotonic environment—one that has the same concentration of dissolved material as the interior of the cell—will experience no net gain or loss of water and will retain its original shape. (b) A cell in a hypotonic environment—one with a lower concentration of dissolved material than the interior of the cell—will gain water and swell. Unlike a bacterial cell, a red blood cell will burst because it lacks a cell wall. (c) A cell in a hypertonic environment—one with a higher concentration of dissolved material than the interior of the cell—will lose water and shrink.

tration 100 percent) toward the water that contains other molecules (concentration less than 100 percent; ➢ Figure 4.30b). Thus, osmosis is the net flow of water molecules from a region of higher concentration of those molecules to a region of lower concentration across a semipermeable membrane (➢ Figure 4.30c).

Osmotic pressure is defined as the pressure required to *prevent* the net flow of water by osmosis. The least amount of hydrostatic pressure required to prevent the movement of water from a given solution into pure water is the osmotic pressure of the solution. The osmotic pressure of a solution is proportional to the number of particles dissolved in a given volume of that solution. Thus, NaCl and other salts that form two ions per molecule exert twice as much osmotic pressure as glucose and other substances that do not ionize, provided each compound is present at the same concentration.

The important thing for a microbiologist to know about osmosis and osmotic pressure is how particles dissolved in fluid environments affect microorganisms in those environments (➢ Figure 4.31). For this purpose, tonicity is a useful concept. *Tonicity* describes the behavior of cells in a fluid environment. The cells are the reference point, and the fluid environments are compared to the cells. The fluid surrounding cells is **isotonic** to the cells when no change in cell volume occurs (➢ Figure 4.31a). The fluid is **hypertonic** to the cells if the cells shrivel or shrink as water moves out of them into the fluid environment (➢ Figure 4.31b); it is **hypotonic** to the cells if the cells swell or burst as water moves from the environment into the cells (➢ Figure 4.31c). Although bacteria become dehydrated and their cytoplasm shrinks away from the cell wall in a hypertonic environment, their cell walls usually prevent them from swelling or bursting in the hypotonic environments they typically inhabit.

Active Transport

In contrast to passive processes, **active transport** moves molecules and ions against concentration gradients from regions of lower concentration to those of higher concentration. This process is analogous to rolling something uphill, and it does require the cell to expend energy from ATP. Active transport is important in microorganisms for moving nutrients that are present in low concentrations in the environment of the cells. It requires membrane proteins that act as both carriers and enzymes (➢ Figure 4.32). These proteins display specificity in that each carrier transports a single substance or a few closely related substances. The results of active transport are to concentrate a substance on one side of a membrane and to maintain that concentration against a gradient. As with facilitated diffusion, active transport carriers also are subject to saturation and competition for binding sites by similar molecules.

Group translocation reactions move a substance from the outside of a bacterial cell to the inside while chemically modifying the substance so that it cannot diffuse out. This process allows molecules such as glucose to be accumulated against a concentration gradient. Because the modified molecule inside the cell is different from those outside, no actual concentration exists. Energy for this process is supplied by phosphoenolpyruvate (PEP), a high-energy phosphate compound. Many eukaryotic cells have a similar active transport mechanism for preventing diffusion.

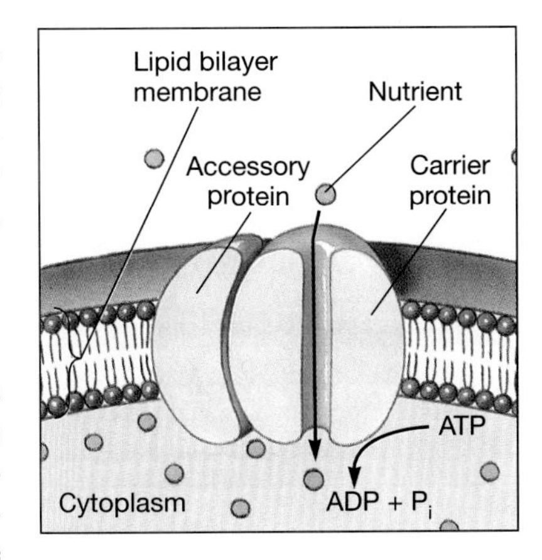

➢ Figure 4.32 **Active transport.** Carrier protein molecules aid in movement of molecules through the membrane. This process can take place against a concentration gradient and so requires the use of energy (in the form of ATP) by the cell. The accessory protein participates in the carrier protein's function. (P_i is inorganic phosphate, HPO_4^{2-}.)

Endocytosis and Exocytosis

In addition to the processes that move substances directly across membranes, eukaryotic cells move substances by forming membrane-enclosed vesicles. Such vesicles are made from portions of the plasma membrane. If they form by invagination (poking in) and surround substances outside the cell, the process is called **endocytosis.** These vesicles pinch off from the plasma membrane and enter the cell. If vesicles inside the cell fuse with the plasma membrane and extrude their contents from the cell, the process is called **exocytosis.** Both endocytosis and exocytosis require energy, probably to allow contractile proteins of the cell's cytoskeleton to move vesicles.

Endocytosis

Several types of endocytosis occur. In one type, known as *receptor-mediated endocytosis,* a substance outside the cell binds to the plasma membrane, which invaginates and surrounds the substance. The exact mechanisms that trigger binding and invagination depend on specific receptor sites on the plasma membrane. Once the substance is completely surrounded by plasma membrane to form a vesicle, the vesicle pinches off from the plasma membrane.

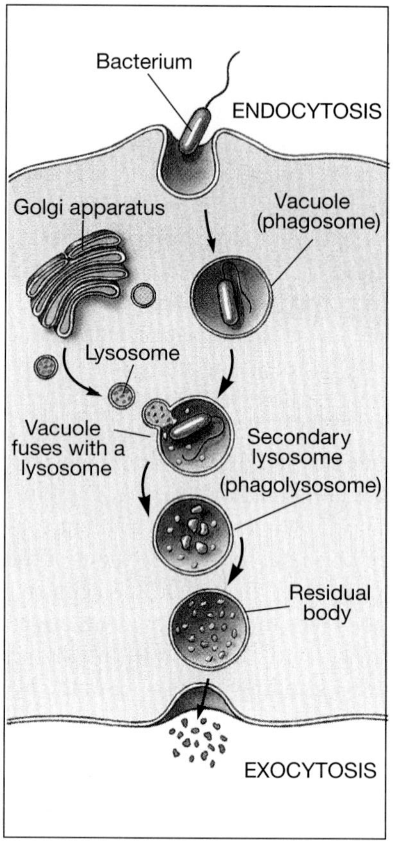

➢ Figure 4.33 **Endocytosis and exocytosis.** Endocytosis is the process of taking materials into the cell; exocytosis is the process of releasing materials from the cell. Material taken in by the form of endocytosis called *phagocytosis* is enclosed in vacuoles known as *phagosomes.* The vacuoles fuse with lysosomes, which release powerful enzymes that degrade the vacuolar contents. Reusable components are absorbed into the cell, and debris is released by exocytosis.

Of all the types of endocytosis, only phagocytosis is of special interest to microbiologists. In **phagocytosis,** large vacuoles called *phagosomes* form around microorganisms and debris from tissue injury. These vacuoles enter the cell, taking with them large amounts of the plasma membrane (➢ Figure 4.33). The vacuole membrane fuses with lysosomes, which release their enzymes into the vacuoles. The enzymes digest the contents of the vacuoles (*phagolysosomes*) and release small molecules into the cytoplasm. Often, undigested particles in residual bodies are returned to and fuse with the plasma membrane. The particles are released from the cell by exocytosis. Certain white blood cells are especially adept at phagocytosis and play an important role in defending the body against infection by microorganisms.

Exocytosis

Exocytosis, the mechanism by which cells release secretions, can be thought of as the opposite of endocytosis. Most secretory products are synthesized on ribosomes or smooth endoplasmic reticulum. They are transported through the membrane of the endoplasmic reticulum, packaged in vesicles, and moved to the Golgi apparatus, where their contents are processed to form the final secretory product. Once secretory vesicles form, they move toward the plasma membrane and fuse with it (➢ Figure 4.33). The contents of the vesicles are then released from the cell.

✓ Most prokaryotes lack sterols in their plasma membranes. What functions do sterols, such as cholesterol, fulfill in eukaryotic plasma membranes?

✓ Compare the number and structure of chromosomes in prokaryotes and eukaryotes.

✓ Give two specific arguments that support the idea that prokaryotes were involved in the evolution of eukaryotes by means of endosymbiosis.

Retracing Our Steps

Basic Cell Types

- Both **prokaryotic cells** and **eukaryotic cells** have membranes that define the bounds of the living cell, and both contain genetic information stored in DNA.
- Prokaryotic cells differ from eukaryotic cells in that they lack a defined nucleus and membrane-enclosed **organelles** (except for a few simple membrane-covered bodies in certain types of prokaryotes).

Prokaryotic Cells

- All prokaryotes are classified in either the domain Archaea or the domain Bacteria.

Size, Shape, and Arrangement

- Prokaryotes are the smallest living organisms.
- Bacteria are grouped by shape: **cocci** (spherical), **bacilli** (rod-shaped), **spirilli** (rigid, wavy), **vibrios** (comma-shaped), and **spirochetes** (corkscrew-shaped).
- Arrangements of bacteria include groupings such as pairs, tetrads, grapelike clusters, and long chains.

An Overview of Structure

- Bacterial cells include a cell membrane, cytoplasm, ribosomes, a nuclear region, and external structures.

The Cell Wall

- The rigid **cell wall** outside the cell membrane is composed mainly of the polymer **peptidoglycan.**
- Cell walls differ in compostition and structure. In Gram-positive bacteria, the cell wall consists of a thick, dense layer of peptidoglycan, with **teichoic acid** in it. In Gram-negative bacteria, the cell wall has a thin layer of peptidoglycan, separated from the cytoplasmic membrane by the **periplasmic space** and enclosed by an **outer membrane** made of **lipopolysaccharide,** or **endotoxin.** In acid-fast bacteria, the cell wall consists mainly of lipids, some of which are true waxes, and some of which are glycolipids.
- Some bacterial cell walls are damaged by penicillin and lysozyme.

The Cell Membrane

- The **cell membrane** has a **fluid-mosaic** structure with phospholipids forming a bilayer and proteins interspersed in a mosaic pattern.
- The main function of the cell membrane is to regulate the movement of materials into and out of cells.
- Bacterial cell membranes also perform functions usually carried out by organelles of eukaryotic cells.

Internal Structure

- The **cytoplasm** is the semifluid substance inside the cell membrane.
- **Ribosomes,** which consist of RNA and protein, serve as sites for protein synthesis.
- The **nuclear region** includes a single, large, circular chromosome, which contains the prokaryotic cell's DNA and some RNA and protein.
- Bacteria contain a variety of **inclusions,** including **granules** that store glycogen or other substances and **vesicles** filled with gas.
- Some bacteria form resistant **endospores.** The core of an endospore contains living material and is surrounded by a cortex, spore coat, and exosporium.

External Structure

- Motile bacteria have one or more **flagella,** which propel the cell by the action of rings in their basal body.
- Much bacterial movement is random, but some bacteria exhibit **chemotaxis** (movement toward attractants and away from repellents) and/or **phototaxis** (movement toward or away from light).
- Some bacteria have **pili**: **Conjugation pili** allow exchange of DNA, whereas **attachment pili** (**fimbriae**) help bacteria adhere to surfaces.
- The **glycocalyx** includes all polysaccharides external to a bacterial cell wall. **Capsules** prevent host cells from destroying a bacterium; capsules of any species of bacteria have a specific chemical composition. **Slime layers** protect bacterial cells from drying, trap nutrients, and sometimes bind cells together, as in dental plaque.

Eukaryotic Cells

An Overview of Structure

- Eukaryotic cells, which are generally larger and more complex than prokaryotic cells, are the basic structural unit of microscopic and macroscopic organisms of the kingdoms Protista, Plantae, Fungi, and Animalia.

The Plasma Membrane

- **Plasma membranes** of eukaryotic cells are almost identical to those of prokaryotic cells, except that they contain sterols. The function of eukaryotic plasma membranes, however, is limited primarily to regulating movement of substances into and out of cells.

Internal Structure

- Eukaryotic cells are characterized by the presence of a membrane-enclosed **cell nucleus,** with a **nuclear envelope, nucleoplasm, nucleoli,** and **chromosomes** (typically paired) that contain DNA and proteins called **histones.**
- In cell division by **mitosis,** each cell receives one of each chromosome found in parent cells. In cell division by **meiosis,** each cell receives one member of each pair of chromosomes, and the progeny can be **gametes** or **spores.**
- **Mitochondria,** the powerhouses of eukaryotic cells, carry out the oxidative reactions that capture energy in ATP.
- Photosynthetic cells contain **chloroplasts,** which capture energy from light.
- Eukaryotic ribosomes are larger than those of prokaryotes and can be free or attached to endoplasmic reticulum. Free ribosomes make protein to be used in the cell; those that are attached to endoplasmic reticulum make proteins to be secreted.
- The **endoplasmic reticulum** is an extensive membrane network. Without ribosomes (smooth ER), the endoplasmic reticulum synthesizes lipids; when combined with ribosomes (rough ER), it produces proteins.
- The **Golgi apparatus** is a set of stacked membranes that receive, modify, and package proteins into **secretory vesicles.**

- **Lysosomes,** in animal cells, are organelles that contain digestive enzymes, which destroy dead cells and digest contents of vacuoles.
- **Peroxisomes** are membrane-enclosed organelles that convert peroxides to water and oxygen and sometimes oxidize amino acids and fats.
- **Vacuoles** contain various stored substances and materials engulfed by phagocytosis.
- The **cytoskeleton** is a network of **microfilaments** and **microtubules** that support and give rigidity to cells and provide for cell movements.

External Structure

- Most external components of eukaryotic cells are concerned with movement. Eukaryotic flagella are composed of microtubules; sliding of proteins at their bases causes them to move.
- **Cilia** are smaller than flagella and beat in coordinated waves.
- **Pseudopodia** are projections into which cytoplasm flows, causing a creeping movement.
- Eukaryotic cells of the plant and fungi kingdoms have cell walls, as do the algal protists.

Evolution by Endosymbiosis

- The **endosymbiont theory** holds that organelles of eukaryotic cells arose from prokaryotes that had been engulfed and survived to develop a symbiotic relationship by living inside the larger cell.
- Mitochondria, chloroplasts, flagella, and microtubules are believed to have originated from endosymbiont prokaryotes.
- Many examples exist of modern prokaryotes living endosymbiotically inside eukaryotes.

The Movement of Substances Across Membranes

- All passive processes involved in movement across membranes involve net movement of substances from a region of higher concentration to a region of lower concentration. These processes do not require expenditure of energy by the cell.

Simple Diffusion

- **Simple diffusion** results from the molecular kinetic energy and random movement of particles. The role of diffusion in living cells depends on the size of particles, nature of membranes, and distances substances must move inside cells.

Facilitated Diffusion

- **Facilitated diffusion** uses protein carrier molecules or protein-lined pores in membranes in moving ions or molecules from high to low concentrations.

Osmosis

- **Osmosis** is the net movement of water molecules through a **selectively permeable** membrane from a region of higher to a region of lower concentration. The **osmotic pressure** of a solution is the pressure required to prevent such a flow.

Active Transport

- Active processes involved in movement of substances across membranes generally result in movement from regions of lower concentration to those of higher concentration and require the cell to expend energy.
- **Active transport** requires a protein carrier molecule in a membrane, a source of ATP, and an enzyme that releases energy from ATP.
- Active transport is important in cell functions because it allows cells to take up substances that are in low concentration in the environment and to concentrate those substances within the cell.

Endocytosis and Exocytosis

- **Endocytosis** and **exocytosis,** which occur only in eukaryotic cells, involve formation of vesicles from fragments of plasma membrane and fusion of vesicles with the plasma membrane, respectively.
- In endocytosis the vesicle enters the cell, as in **phagocytosis**.
- In exocytosis the vesicle leaves the cell, as in secretion.
- Endocytosis and exocytosis are important because they allow the movement of relatively large quantities of materials across plasma membranes.

Terminology Check

active transport (*p. 101*)
amoeboid movement (*p. 95*)
amphitrichous (*p. 85*)
atrichous (*p. 85*)
attachment pilus (*p. 89*)
axial filament (*p. 88*)
bacillus (*p. 73*)
capsule (*p. 89*)
cell membrane (*p. 80*)
cell nucleus (*p. 91*)
cell wall (*p. 75*)
chemotaxis (*p. 88*)
chloroplast (*p. 92*)
chromatin (*p. 92*)
chromatophore (*p. 82*)
chromosome (*p. 92*)
cilium (*p. 94*)
coccus (*p. 73*)
conjugation pilus (*p. 89*)
crista (*p. 92*)
cytoplasm (*p. 82*)
cytoplasmic streaming (*p. 95*)
cytoskeleton (*p. 94*)
diploid (*p. 92*)
dyad (*p. 92*)
endocytosis (*p. 101*)
endoflagellum (*p. 88*)
endoplasmic reticulum (*p. 93*)
endospore (*p. 83*)
endosymbiotic theory (*p. 96*)
endotoxin (*p. 77*)
eukaryotic cell (*p. 71*)
exocytosis (*p. 101*)
facilitated diffusion (*p. 99*)
fimbria (*p. 89*)
flagellum (*p. 85*)
fluid-mosaic model (*p. 80*)
gamete (*p. 92*)
glycocalyx (*p. 89*)
Golgi apparatus (*p. 93*)
granule (*p. 83*)
group translocation reaction (*p. 101*)
haploid (*p. 92*)
histone (*p. 92*)
hydrophilic (*p. 80*)
hydrophobic (*p. 80*)
hypertonic (*p. 101*)
hypotonic (*p. 101*)

inclusion (*p. 83*)
isotonic (*p. 101*)
lipid A (*p. 77*)
lipopolysaccharide (LPS) (*p. 77*)
lophotrichous (*p. 85*)
lysosome (*p. 93*)
matrix (*p. 92*)
meiosis (*p. 92*)
metachromasia (*p. 83*)
metachromatic granule (*p. 83*)
microfilament (*p. 94*)
microtubule (*p. 94*)
mitochondrion (*p. 92*)
mitosis (*p. 92*)
monotrichous (*p. 85*)
nuclear envelope (*p. 91*)
nuclear pore (*p. 92*)
nuclear region (*p. 82*)
nucleoid (*p. 82*)
nucleolus (*p. 92*)
nucleoplasm (*p. 92*)
organelle (*p. 71*)
osmosis (*p. 99*)
osmotic pressure (*p. 101*)
outer membrane (*p. 77*)
pellicle (*p. 89*)
peptidoglycan (*p. 75*)
periplasmic space (*p. 77*)
peritrichous (*p. 85*)
peroxisome (*p. 94*)
phagocytosis (*p. 102*)
phototaxis (*p. 88*)
pilus (*p. 88*)
plasma membrane (*p. 90*)
pleomorphism (*p. 73*)
polyribosome (*p. 82*)
prokaryotic cell (*p. 71*)
protoplast (*p. 78*)
pseudopodium (*p. 95*)
ribosome (*p. 82*)
secretory vesicle (*p. 93*)
selectively permeable (*p. 99*)
simple diffusion (*p. 98*)
slime layer (*p. 89*)
spheroplast (*p. 78*)
spindle apparatus (*p. 92*)
spirillum (*p. 73*)
spirochete (*p. 73*)
spore (*p. 92*)
stroma (*p. 92*)
teichoic acid (*p. 77*)
thylakoid (*p. 92*)
vacuole (*p. 94*)
vegetative cell (*p. 83*)
vesicle (*p. 83*)
vibrio (*p. 73*)
volutin (*p. 83*)
zygote (*p. 92*)

See You on the Web

Don't forget I am expecting you on the web. The following is a preparedness checklist for your web explorations. If you have questions on any of these items, you can return to the end of Chapter 1 for more details.

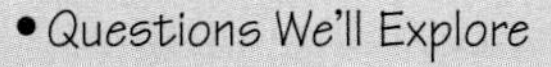

- Questions We'll Explore
- Concept Check questions
- Flashcards
- Study Guide

If you are ready, come visit me at

http://www.wiley.com/college/black

It will be a journey you'll never forget.

Essential Concepts of Metabolism

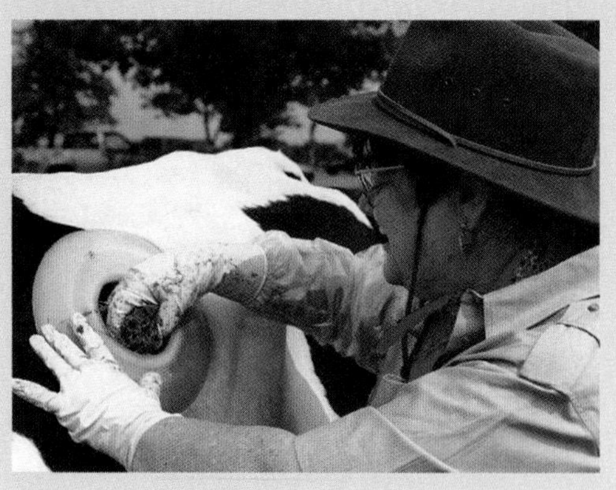

You're right! I do have my arm inside the cow. It's nice and warm and squishy there inside her rumen, one of the cow's four stomach-like compartments. The grass and hay she's eaten are being digested there. Why can't we digest such low-cost meals? We lack the enzymes needed for the metabolic pathways that digest grass—as does the cow. However, she has billions of microbes, a different mix in each of the four "stomachs," that do metabolize the grass for her. Without them she would starve. I'm going to remove a sample of rumen contents, squeeze out the juice, and examine it under the microscope, plus try to grow some of the fascinating microbes in culture. Microbes are able to do far more types of metabolism than are humans. And remember; most of the methane in the earth's atmosphere comes from rumen microbes. Microbial metabolism keeps our world running.

Background Basics

molecular structure . . . . *(Chapter 2)*
bonding . . . . *(Chapter 2)*
ionization . . . . *(Chapter 2)*
oxidation, reduction . . . . *(Chapter 2)*
ATP and energy . . . . *(Chapter 2)*
building blocks of carbohydrates, lipids, proteins . . . *(Chapter 2)*
peptide bonds, amino acids, amino group . . . . *(Chapter 2)*
protein structure, denaturing . . . . *(Chapter 2)*
pH . . . . *(Chapter 2)*
cell wall . . . . *(Chapter 4)*
cell membranes . . . . *(Chapter 4)*
flagella: structure and action . . . . *(Chapter 4)*
spirochete endoflagella . . . . *(Chapter 4)*

Questions We'll Explore

A How do the following terms relate to metabolism: autotrophy, heterotrophy, oxidation, reduction, photoautotrophy, photoheterotrophy, chemoautotrophy, chemoheterotrophy, glycolysis, fermentation, aerobic metabolism, and biosynthetic processes?

B What are the characteristics of enzymes, and how do those characteristics contribute to their function?

C What are the main steps and significance of glycolysis and fermentation?

D What are the main steps and significance of the Krebs cycle?

E What are the roles of electron transport and oxidative phosphorylation in energy capture?

F How do microorganisms metabolize fats and proteins for energy?

G What are the main steps and significance of photosynthesis in microbes?

H How do photoheterotrophy and chemoautotrophy differ?

I How do bacteria carry out biosynthetic activities?

J How do bacteria use energy for membrane transport and for movement?

Until the middle of the nineteenth century, people didn't know what caused a fruit juice to become wine, or milk to sour. Then, in 1857, Louis Pasteur proved that alcoholic fermentation was due to microorganisms. A few years later he identified specific organisms from samples of fermenting juices and souring milk. Pasteur was one of the first to study chemical processes in a living organism. Since his time, much has been learned about such processes.

Metabolism: An Overview

Metabolism is the sum of all the chemical processes carried out by living organisms (➢ Figure 5.1). It includes **anabolism,** reactions that require energy to synthesize complex molecules from simpler ones, and **catabolism,** reactions that release energy by breaking complex molecules into simpler ones that can then be reused as building blocks. Anabolism is needed for growth, reproduction, and repair of cellular structures. Catabolism provides the organism with energy for its life processes, including movement, transport, and the synthesis of complex molecules—that is, anabolism.

All catabolic reactions involve *electron transfer,* which allows energy to be captured in high-energy bonds in ATP and similar molecules. ➡ (Appendix E) Electron transfer is directly related to oxidation and reduction (Table 5.1). **Oxidation** can be defined as the loss or removal of electrons. Although many substances combine with

Membrane Transport and Movement

In addition to using energy for biosynthetic processes, microorganisms also use energy for transporting substances across membranes and for their own movement. These energy uses are as important to the survival of the organisms as are their biosynthetic activities.

Membrane Transport

Microbes use energy to move most ions and metabolites across cell membranes against concentration gradients. For example, bacteria can transport a sugar or an amino acid from a region of low concentration outside the cell to a region of higher concentration inside the cell. This means that they accumulate nutrients within cells in concentrations a hundred to a thousand times the concentration outside the cell. They also concentrate certain inorganic ions by the same means.

Two mechanisms exist in bacteria for concentrating substances inside cells, and both require energy. One active transport mechanism is specific to Gram-negative bacteria, such as *E. coli.* Such bacteria have two membranes—the cell membrane, which surrounds the cell's cytoplasm, and the outer membrane, which forms part of the cell wall. ➡ (Chapter 4, p. 75) Transmembrane carrier proteins called **porins** form channels through the outer membrane. Porins allow entry of ions and small hydrophilic metabolites via *facilitated diffusion.* ➡ (Chapter 4, p. 99) After entering the periplasmic space, a specific periplasmic protein combines with one of the diffusing ions or metabolites. The periplasmic protein then facilitates the transport of the substance into the cytoplasm via a specific carrier protein in the cell membrane. Such substances generally gain entry by active transport. Through ATP hydrolysis, the carrier protein changes shape, allowing the metabolite into the cytoplasm (see Figure 4.32).

Another mechanism, present in all bacteria, is called the **phosphotransferase system (PTS).** It consists of sugar-specific enzyme complexes called **permeases** (per'me-a-sez), which form a transport system through the cell membrane. The PTS uses energy from the high-energy phosphate molecule phosphoenolpyruvate (PEP). When PEP is present in the cytoplasm, it can provide energy and a phosphate group to a permease in the membrane. Then the permease transfers the phosphate to a sugar molecule and at the same time moves the sugar across the membrane. A phosphorylated sugar is thus transported inside the cell and is prepared to undergo metabolism. This *group translocation* was discussed in Chapter 4. ➡ (p. 101)

Movement

Most motile bacteria move by means of flagella, but some move by gliding or creeping or in a corkscrew motion. Flagellated bacteria move by rotating their flagella. ➡ (Chapter 4, p. 85) The mechanism for rotation, though not fully understood, appears to involve a proton gradient as in chemiosmosis. As the protons move down the gradient, they drive the rotation. Gliding bacteria move only when in contact with a solid surface, such as decaying organic matter. Rotation of the cell on its own axis often occurs with gliding. A number of mechanisms have been proposed to explain gliding, but the mechanism that propels the gliding bacterium *Myxococcus* is best understood. This organism uses energy to secrete a substance called a **surfactant** (ser-fak'tant), which lowers surface tension at the bacterium's posterior end. The difference in surface tension between the anterior and posterior ends (a passive phenomenon) causes *Myxococcus* to glide.

Spirochetes expend energy for both creeping and thrashing motions. On a solid surface they creep along like an inchworm by alternately attaching front and rear ends. Suspended in a liquid medium, they thrash (twist and turn). Both creeping and thrashing motions probably occur by waves of contraction within the cell substance that exert force against axial filaments.

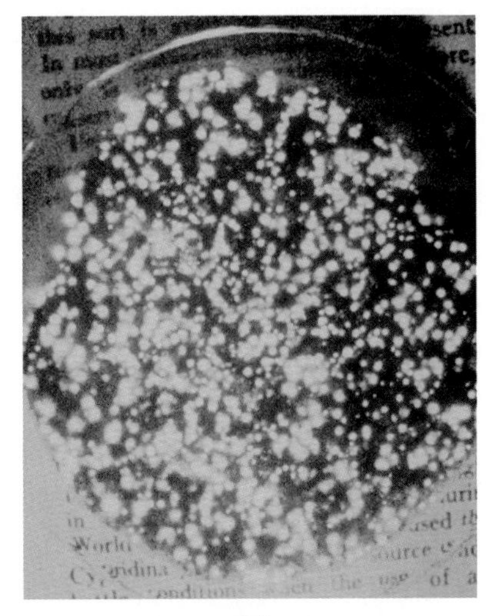

(a)

(b)

➢ Figure 5.28 **Microbial bioluminescence.** (a) Bioluminescent bacteria in the Petri dish produce enough light to read by. (b) Angler fish lights up the dark deep-ocean depths with bioluminescent bacteria living symbiotically in its long "lure" that attracts prey to within reach of its jaws.

Bioluminescence

Bioluminescence, the ability of an organism to emit light, appears to have evolved as a byproduct of aerobic metabolism. Bacteria of the genera *Photobacterium* and *Achromobacter,* fireflies, glowworms, and certain marine organisms living at great depths in the ocean exhibit bioluminescence (➢ Figure 5.28). Many light-emitting organisms have the enzyme *luciferase* (lu-sif'er-ace), along with other components of the electron transport system. (Luciferase derives its name from Lucifer, which means "morning star.") Luciferase catalyzes a complex reaction in which molecular oxygen is used to oxidize a long-chain aldehyde or ketone to a carboxylic acid. At the same time, $FMNH_2$ from the electron transport chain is oxidized to an excited form of *flavin mononucleotide* (FMN), a carrier molecule derived from riboflavin (vitamin B_2), that emits light as it returns to its unexcited state. In this process, phosphorylation reactions are bypassed, and no ATP is generated. Instead, energy is released as light.

Luminescent microorganisms often live on the surface of marine organisms such as some squids and fish. More than 300 years ago, the Irish chemist Robert Boyle observed that the familiar glow of the skin of dead fish lasted only as long as oxygen was available. At that time the electron transport system and the role of oxygen in it were not understood.

Bioluminescence exhibited by larger organisms has survival value. It is the sole light source for marine creatures that live at great depths, and it helps land organisms such as fireflies find mates. How bioluminescence came to be established among microorganisms is less clear. One hypothesis is that early in the evolution of living things, bioluminescence served to remove oxygen from the atmosphere as it was produced by some of the first photosynthetic organisms. Although this is not an advantage to aerobes, it is an advantage to strict anaerobes. Because most of the microorganisms in existence at that time were anaerobes susceptible to the toxic effects of oxygen, bioluminescence would have been beneficial to them. Today, many bioluminescent microbes are beneficiaries of symbiotic relationships with their hosts. They provide light in return for a shelter and nutrients.

Scientists have found a way to put bioluminescent bacteria to work. In the Microtox Acute Toxicity Test, bioluminescent bacteria are exposed to a water sample to determine if the sample is toxic. Any change—positive or negative—in the bacteria's growth is observable as a change in their light output. The sample's toxicity is calculated by comparing before-and-after readings of the light levels. The brainchild of Microbics Corp. of Carlsbad, California, this toxicity test takes only minutes to perform.

The Microtox Acute Toxicity Test is useful for testing the quality of drinking water and for numerous other industrial applications. For example, waste-water treatment plants use it to determine quickly whether their treated effluent will be able to pass government toxicity compliance tests. Paper mills use the test to determine how much disinfectant is needed to rid their equipment of the microbial growth that slows down the manufacturing process and affects product quality. Makers of household cleansers, shampoos, or cosmetics use the test in place of controversial animal testing in which drops of the products are put into the eyes of rabbits to determine the products' irritancy levels. And unlike cell-culturing techniques, the test requires little skill to perform and to interpret. Bioluminescence could prove to be a very important process to industry in the future.

Retracing Our Steps

Metabolism: An Overview

- **Metabolism** is the sum of all the chemical processes in a living organism. It consists of **anabolism,** reactions that require energy to synthesize complex molecules from simpler ones, and **catabolism,** reactions that release energy by breaking complex molecules into simpler ones.
- **Autotrophs,** which use carbon dioxide to synthesize organic molecules, include **photoautotrophs** (which carry on photosynthesis) and **chemoautotrophs.**
- **Heterotrophs,** which use organic molecules made by other organisms, include **chemoheterotrophs** and **photoheterotrophs.**
- For growth, movement, and other activities, **metabolic pathways** use energy captured in the **catabolic pathways.**

Enzymes

Properties of Enzymes

- **Enzymes** are proteins that catalyze chemical reactions in living organisms by lowering the **activation energy** needed for a reaction to occur.
- Enzymes have an **active site,** the binding site to which the **substrate** (the substance on which the enzyme acts) attaches to form an **enzyme–substrate complex**. Enzymes typically exhibit a high degree of **specificity** in the reactions they catalyze.

Properties of Coenzymes and Cofactors

- Some enzymes require **coenzymes,** nonprotein organic molecules that can combine with the **apoenzyme,** the protein portion of the enzyme, to form a **holoenzyme.** Some enzymes also require inorganic ions as **cofactors.**

Enzyme Inhibition

- Enzyme activity can be reduced by **competitive inhibitors,** molecules that compete with the substrate for the enzyme's active site, or by **noncompetitive inhibitors,** molecules that bind to an **allosteric site,** a site other than the active site.

Factors That Affect Enzyme Reactions

- Factors that affect the rate of enzyme reactions include temperature, pH, and concentrations of substrate, product, and enzyme.

Anaerobic Metabolism: Glycolysis and Fermentation

Glycolysis

- **Glycolysis** is a metabolic pathway by which glucose is oxidized to pyruvic acid.
- Under anaerobic conditions, glycolysis yields a net of two ATPs per molecule of glucose.

Fermentation

- **Fermentation** refers to the reactions of metabolic pathways by which NADH is oxidized to NAD.
- Six pathways of fermentation are summarized in Figure 5.12. **Homolactic acid** and **alcoholic fermentations** are two of the most important and commonly occurring fermentation pathways.

Aerobic Metabolism: Respiration

- **Anaerobes** do not use oxygen; **aerobes** use oxygen and obtain energy chiefly via **aerobic respiration.**

The Krebs Cycle

- The **Krebs cycle** involves the metabolism of two-carbon compounds to CO_2 and H_2O, the production of one ATP directly from each acetyl group, and the transfer of hydrogen atoms to the electron transport system.
- In energy production the Krebs cycle processes acetyl-CoA so that (in the electron transport chain) hydrogen atoms can be oxidized for energy.

Electron Transport and Oxidative Phosphorylation

- **Electron transport** is the transfer of electrons to oxygen (the final electron acceptor).
- **Oxidative phosphorylation** involves the **electron transport chain** for ATP synthesis and is a membrane-regulated process not directly related to the metabolism of specific substrates.
- The theory of **chemiosmosis** explains how energy is used to synthesize ATP.

The Significance of Energy Capture

- In prokaryotes, aerobic (oxidative) metabolism captures 19 times as much energy as does anaerobic metabolism.

The Metabolism of Fats and Proteins

- Most organisms get energy mainly from glucose. But for almost any organic substance, there is some microorganism that can metabolize it.

Fat Metabolism

- Fat metabolism involves hydrolysis and the enzymatic formation of glycerol and free fatty acids. Fatty acids are in turn oxidized by **beta oxidation,** which results in the release of acetyl-CoA. Acetyl-CoA then enters the Krebs cycle.

Protein Metabolism

- The metabolism of proteins involves the breakdown of proteins to amino acids, the deamination of the amino acids, and their subsequent metabolism in glycolysis, fermentation, or the Krebs cycle.

Other Metabolic Processes

Photoautotrophy

- **Photosynthesis** is the use of light energy to synthesize carbohydrates: (1) The **light reactions** can involve **cyclic photophosphorylation** or **photolysis** accompanied by **noncyclic photoreduction** of NADP; (2) the **dark reactions** involve the reduction of CO_2 to carbohydrate.
- Photosynthesis in cyanobacteria and algae provides a means of making nutrients, as it does in green plants; however, photosynthetic bacteria generally use some substances besides water to reduce carbon dioxide.

Photoheterotrophy

- **Photoheterotrophy** is the use of light as a source of energy. It requires organic compounds as sources of carbon.

Chemoautotrophy

- **Chemoautotrophs,** or **chemolithotrophs,** oxidize inorganic substances to obtain energy. Chemolithotrophs require only carbon dioxide as a carbon source.

The Uses of Energy

Biosynthetic Activities

- An **amphibolic pathway** is a metabolic pathway that can capture energy or synthesize substances needed by the cell.
- Figure 5.26 summarizes the intermediate products of energy-yielding metabolism and some of the building blocks for synthetic reactions that can be made from them.
- Bacteria synthesize a variety of cell wall polymers.

Membrane Transport and Movement

- Membrane transport uses energy derived from the ATP-producing electron transport system in the membrane to concentrate substances against a gradient. It occurs by active transport and by the **phosphotransferase system.**
- Movement in bacteria can be by flagella, by gliding or creeping, or by axial filaments.

Terminology Check

activation energy (*p. 109*)
active site (*p. 110*)
aerobe (*p. 119*)
aerobic respiration (*p. 119*)
alcoholic fermentation (*p. 118*)
allosteric site (*p. 112*)
amphibolic pathway (*p. 131*)
anabolic pathway (*p. 109*)
anabolism (*p. 106*)
anaerobe (*p. 119*)
apoenzyme (*p. 110*)
autotroph (*p. 107*)
autotrophy (*p. 107*)
beta oxidation (*p. 125*)
catabolic pathway (*p. 109*)
catabolism (*p. 106*)
chemical equilibrium (*p. 114*)
chemiosmosis (*p. 123*)
chemoautotroph (*p. 107*)
chemoheterotroph (*p. 108*)
citric acid cycle (*p. 119*)
coenzyme (*p. 110*)
cofactor (*p. 111*)
competitive inhibitor (*p. 112*)
cyclic photophosphorylation (*p. 128*)
cytochrome (*p. 122*)
dark reactions (*p. 129*)
electron acceptor (*p. 107*)
electron donor (*p. 107*)
electron transport (*p. 121*)
electron transport chain (*p. 121*)
endoenzyme (*p. 110*)
enzyme (*p. 109*)
enzyme–substrate complex (*p. 110*)
exoenzyme (*p. 110*)
FAD (*p. 112*)
feedback inhibition (*p. 113*)
fermentation (*p. 116*)
flavoprotein (*p. 122*)
glycolysis (*p. 114*)
heterotroph (*p. 107*)
heterotrophy (*p. 107*)
holoenzyme (*p. 110*)
homolactic acid fermentation (*p. 117*)
Krebs cycle (*p. 119*)
light reactions (*p. 128*)
metabolic pathway (*p. 109*)
metabolism (*p. 106*)
NAD (*p. 110*)
noncompetitive inhibitor (*p. 112*)
noncyclic photoreduction (*p. 129*)
oxidation (*p. 106*)
oxidative phosphorylation (*p. 122*)
permease (*p. 133*)
phosphorylation (*p. 116*)
phosphotransferase system (PTS) (*p. 133*)
photoautotroph (*p. 107*)
photoheterotroph (*p. 107*)
photolysis (*p. 129*)
photosynthesis (*p. 127*)
porin (*p. 133*)
quinone (*p. 123*)
reduction (*p. 107*)
specificity (*p. 110*)
substrate (*p. 110*)
surfactant (*p. 133*)
tricarboxylic acid (TCA) cycle (*p. 119*)

See You on the Web

Don't forget I am expecting you on the web. The following is a preparedness checklist for your web explorations. If you have questions on any of these items, you can return to the end of Chapter 1 for more details.

- Questions We'll Explore
- Concept Check questions
- Flashcards
- Study Guide

If you are ready, come visit me at

http://www.wiley.com/college/black

It will be a journey you'll never forget.

Growth and Culturing of Bacteria

I stood in awe near the geyser in Iceland, completely oblivious to the grandeur of my surroundings. The entire reason for my expedition undulated gently in the current of the runoff stream before me. The long, waving filaments of sulfur bacteria looked like long blond hair blowing in a light breeze. They were magnificent! Finally I was able to see with my own eyes, bacteria I had been reading about for years.

My excitement overcame me, and despite the warning steam billowing off the water, I plunged my hand into the water. I just wanted to find out what the strands felt like, but I guess I will never know. The near boiling water scalded my hand immediately. Later, as I nursed my blisters and wounded pride, I pondered the phenomena that allows the bacteria to thrive in an environment that is so hostile to most life forms (including me).

Background Basics

arrangements of bacteria ....*(Figure 4.2)*
math tool: logarithm ....*(Appendix A)*
pH ....*(Chapter 2)*
Celsius vs. Fahrenheit ....*(Appendix A)*
enzymes and coenzymes ....*(Chapter 5)*
osmotic relationships ....*(Figure 4.28)*

Questions We'll Explore

A How is growth defined in bacteria?

B How does cell division occur in microorganisms?

C What are the phases of growth in a bacterial culture?

D How is bacterial growth measured?

E How do physical factors affect bacterial growth?

F How do biochemical factors affect bacterial growth?

G What occurs in sporulation, and what is its significance?

H What methods are used to obtain a pure culture of an organism for study in the laboratory?

I How are different nutritional requirements supplied by varous media?

In this chapter, we will use what we learned in Chapter 5 about energy in microorganisms to study how to grow them in the laboratory. Bacterial growth, which has been more thoroughly studied than growth in other microorganisms, is affected by a variety of physical and nutritional factors. Knowing how these factors influence growth is useful in culturing organisms in the laboratory and in preventing their growth in undesirable places. Furthermore, growing the microbes in pure cultures is essential in performing diagnostic tests that are used to identify a number of disease-causing organisms.

Growth and Cell Division

Microbial Growth Defined

In everyday language, growth refers to an increase in size. We are accustomed to seeing children, other animals, and plants grow. Unicellular organisms also grow, but as soon as a cell, called the **mother** (or *parent*) **cell,** has approximately doubled in size and duplicated its contents, it divides into two **daughter cells.** Then the daughter cells grow, and subsequently they also divide. Because individual cells grow larger only to divide into two new individuals, **microbial growth** is defined not in terms of cell size but as the increase in the number of cells, which occurs by cell division.

> Each square centimeter of skin hosts an average of 100,000 organisms. Bacteria reproduce so quickly, their population is restored within hours of washing.

Cell Division

Cell division in bacteria, unlike cell division in eukaryotes, usually occurs by *binary fission* or sometimes by *budding*. In **binary fission,** a cell duplicates its components and divides into two cells (➢ Figure 6.1a). The daughter cells become independent when a *septum* (partition) grows between them and they separate (➢ Figure 6.1b). Unlike eukaryotic cells, prokaryotic cells do not have a cell cycle with a specific period

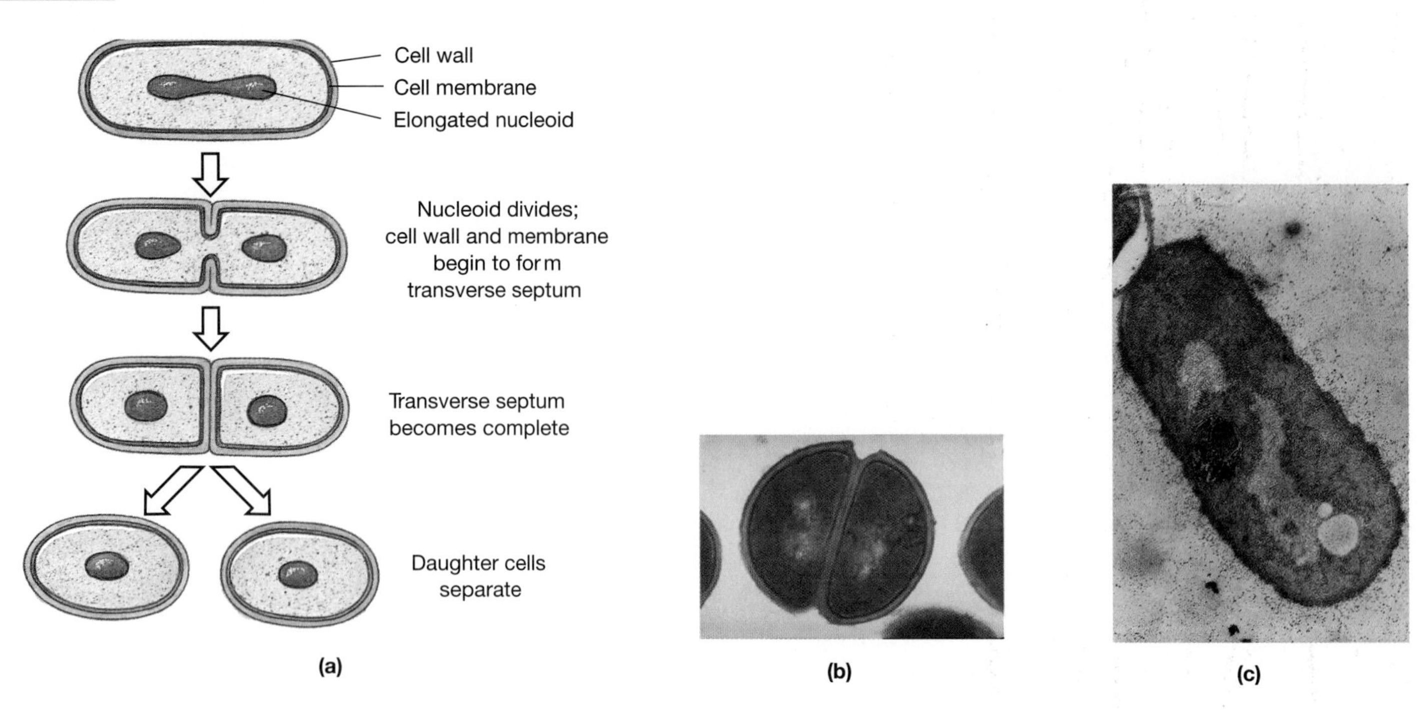

➢ Figure 6.1 **Binary fission.** (a) The stages of binary fission in a bacterial cell. (b) A colorized TEM of a thin section of the bacterium *Staphylococcus* undergoing binary fission. (c) The nucleoid of a bacterial cell.

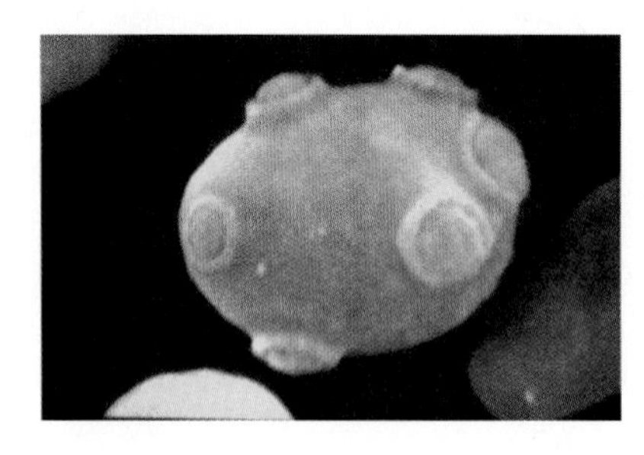

➢ Figure 6.2 **Budding in yeast.**

of DNA synthesis. Instead, in continuously dividing cells, DNA synthesis also is continuous and replicates the single bacterial chromosome shortly before the cell divides. The chromosome is attached to the cell membrane, which grows and separates the replicated chromosomes. Replication of the chromosome is completed before cell division, when the cell may temporarily contain two or more nucleoids. In some species, incomplete separation of the cells produces linear chains (linked bacilli), **tetrads** (cuboidal groups of four cocci), **sarcinae** (singular: *sarcina;* groups of eight cocci in a cubical packet), or grapelike clusters (staphylococci) ➡ (➢ Figure 4.2). Some bacilli always form chains or filaments; others form them only under unfavorable growth conditions. Streptococci form chains when grown on artificial media but exist as single or paired cells when isolated from a rapidly growing lesion in an infected human host.

Cell division in yeast and a few bacteria occurs through **budding.** In that process, a small, new cell develops from the surface of an existing cell and subsequently separates from the parent cell (➢ Figure 6.2).

Phases of Growth

Consider a population of organisms introduced into a fresh, nutrient-rich **medium** (plural: *media*), a mixture of substances on or in which microorganisms grow. Such organisms display four major phases of growth: (1) the lag phase, (2) the log (logarithmic) phase, (3) the stationary phase, and (4) the decline phase, or death phase. These phases form the **standard bacterial growth curve** (➢ Figure 6.3).

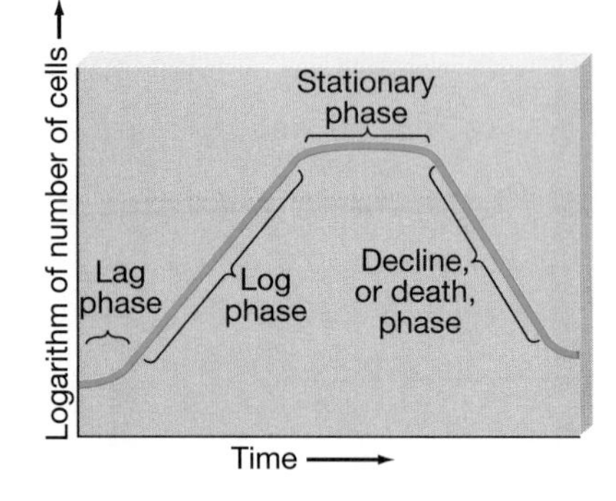

➢ Figure 6.3 **A standard bacterial growth curve.**

The Lag Phase

In the **lag phase,** the organisms do not increase significantly in number, but they are metabolically active—growing in size, synthesizing enzymes, and incorporating various molecules from the medium. During this phase the individual organisms increase in size, and they produce large quantities of energy in the form of ATP.

The length of the lag phase is determined in part by characteristics of the bacterial species and in part by conditions in the media—both the medium from which the organisms are taken and the one to which they are transferred. Some species adapt to the new medium in an hour or two; others take several days. Organisms from old cultures, adapted to limited nutrients and large accumulations of wastes, take longer to adjust to a new medium than do those transferred from a relatively fresh, nutrient-rich medium.

Under ideal conditions, one bacterium can multiply to 2,097,152 within 7 hours.

The Log Phase

Once organisms have adapted to a medium, population growth occurs at an **exponential,** or **logarithmic** (log), **rate.** When the scale of the vertical axis is logarithmic, growth in this **log phase** appears on a graph as a straight diagonal line, which represents the size of the bacterial population. (On the base-10 logarithmic scale, each successive unit represents a 10-fold increase in the number of organisms.) ➡ (Appendix A) During the log phase, the organisms divide at their most rapid rate—a regular, genetically determined interval called the **generation time.** The population of organisms doubles in each generation time. For example, a culture containing 1000 organisms per milliliter with a generation time of 20 minutes would contain 2000 organisms per milliliter after 20 minutes, 4000 organisms after 40 minutes, 8000 after 1 hour, 64,000 after 2 hours, and 512,000 after 3 hours. Such growth is said to be *exponential,* or *logarithmic.*

The generation time for most bacteria is between 20 minutes and 20 hours and is typically less than 1 hour. Some bacteria, such as those that cause tuberculosis and leprosy, have much longer generation times. Some organisms take slightly longer than others to go from the lag phase to the log phase, and they do not all divide precisely together. If they divided together and the generation time was exactly 20 minutes, the number of cells in a culture would increase in a stair-step pattern, exactly doubling every 20 minutes—a hypothetical situation called **synchronous growth.** In an actual culture, each cell divides sometime during the 20-minute generation time, with about 1/20 of the cells dividing each minute—a natural situation called **nonsynchronous growth.** Nonsynchronous growth appears as a straight line, not as steps, on a logarithmic graph (➢ Figure 6.4).

Organisms in a tube of culture medium can maintain logarithmic growth for only a limited period of time. As the number of organisms increases, nutrients are used up, metabolic wastes accumulate, living space may become limited, and aerobes suffer from oxygen depletion. Generally, the limiting factor for logarithmic growth seems to be the rate at which energy can be produced in the form of ATP. As the availability of nutrients decreases, the cells become less able to generate ATP, and their growth rate decreases. The decrease in growth rate is shown in ➢ Figure 6.3 by a gradual leveling off of the growth curve (the curved segment to the right of the log phase).

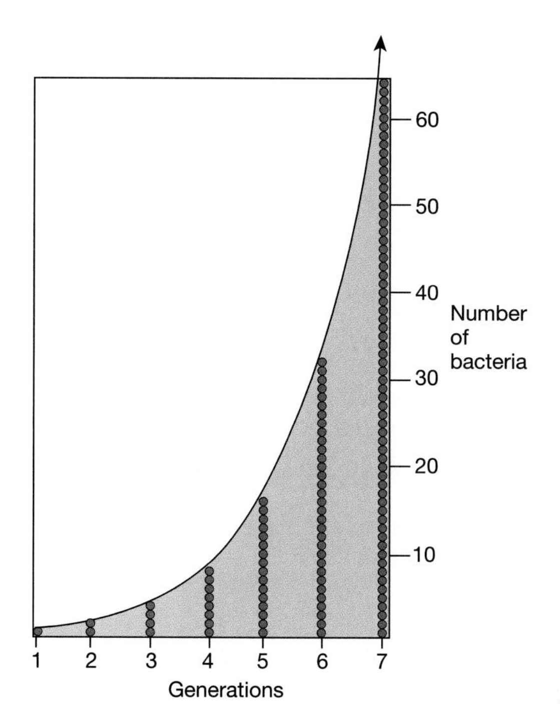

➢ Figure 6.4 **Nonsynchronous growth.** A growth curve for an exponentially increasing population.

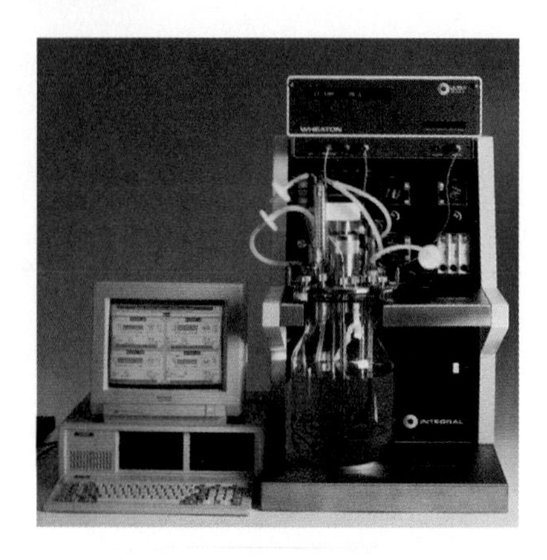

➢ Figure 6.5 **Microbes growing in a chemostat.** A chemostat constantly renews nutrients in a culture, making it possible to grow organisms continuously in the log phase.

Leveling off of growth is followed by the stationary phase unless fresh medium is added or organisms are transferred to fresh medium. Logarithmic growth can be maintained by a device, much like a thermostat, called a **chemostat** (➢ Figure 6.5), which has a growth chamber and a reservoir from which fresh medium is continuously added to the growth chamber as old medium is withdrawn. Alternatively, organisms from a culture in the stationary phase can be transferred to a fresh medium. After a brief lag phase, such organisms quickly reenter the log phase of growth.

The Stationary Phase

When cell division decreases to the point that new cells are produced at the same rate as old cells die, the number of live cells stays constant. The culture is then in the **stationary phase,** represented by a horizontal straight line in ➢ Figure 6.3. The medium contains a limited amount of nutrients and may contain toxic quantities of waste materials. Also, the oxygen supply may become inadequate for aerobic organisms, and damaging pH changes may occur.

The Decline (Death) Phase

As conditions in the medium become less and less supportive of cell division, many cells lose their ability to divide, and thus the cells die. In this **decline phase,** or **death phase,** the number of live cells decreases at a logarithmic rate, as indicated by the straight, downward-sloping diagonal line in ➢ Figure 6.3. During the decline phase, many cells undergo *involution*—that is, they assume a variety of unusual shapes, which makes them difficult to identify. In cultures of spore-forming organisms, more spores than vegetative (metabolically active) cells survive. The duration of this phase is as highly variable as the duration of the logarithmic growth phase. Both depend primarily on the genetic characteristics of the organism. Cultures of some bacteria go through all growth phases and die in a few days; others contain a few live organisms after months or even years.

Growth in Colonies

Growth phases are displayed in different ways in colonies growing on a solid medium. Typically, a cell divides exponentially, forming a small **colony**—all the descendants of the original cell. The colony grows rapidly at its edges; cells nearer the center grow more slowly or begin to die because they have smaller quantities of available nutrients and are exposed to more toxic waste products. All phases of the growth curve occur simultaneously in a colony.

Measuring Bacterial Growth

Bacterial growth is measured by estimating the number of cells that have arisen by binary fission during a growth phase. This measurement is expressed as the number of *viable* (living) organisms per milliliter of culture. Several methods of measuring bacterial growth are available.

Serial Dilution and Standard Plate Counts

One method of measuring bacterial growth is the *standard plate count.* This technique relies on the fact that under proper conditions, only a living bacterium will divide and form a visible colony on an agar plate. An *agar plate* is a nutrient medium solidified with **agar,** a complex polysaccharide extracted from certain marine algae. Because it is difficult to count more than 300 colonies on one agar plate, it is usually necessary to dilute the original bacterial culture before you plate (transfer) a known volume of the culture onto the solid plate. *Serial dilutions* accomplish this purpose.

To make **serial dilutions** (➢ Figure 6.6), you start with organisms in liquid medium. Adding 1 ml of this medium to 9 ml of sterile water makes a 1:10 dilution; adding 1 ml of the 1:10 dilution to 9 ml of sterile water makes a 1:100 dilution; and so on. The number of bacteria per milliliter of fluid is reduced by 9/10 in each dilution. Subsequent dilutions are made in ratios of 1:1000, 1:10,000, 1:100,000, 1:1,000,000, or even 1:10,000,000 if the original culture contained an extremely large number of organisms.